AF784078

Dear Colonel Kelley.

I hope the book will be a useful reference for you.

Ralph A Scott, Jr.

ACS SYMPOSIUM SERIES **345**

Design Considerations for Toxic Chemical and Explosives Facilities

Ralph A. Scott, Jr., EDITOR
Defense Explosives Safety Board

Laurence J. Doemeny, EDITOR
National Institute for Occupational Safety and Health

Developed from a symposium sponsored by
the Division of Chemical Health and Safety
at the 194th Meeting
of the American Chemical Society,
New Orleans, Louisiana
August 30–September 4, 1987

American Chemical Society, Washington, DC 1987

Library of Congress Cataloging-in-Publication Data

Design considerations for toxic chemical and explosives facilities.
(ACS symposium series; 345)

Includes bibliographies and indexes.

1. Explosives—Safety measures—Congresses.
2. Chemicals—Safety measures—Congresses.

I. Scott, Ralph A., 1930- . II. Doemeny, Laurence J. III. American Chemical Society. Division of Chemical Health and Safety. IV. American Chemical Society. Meeting (194th: 1987: New Orleans, La.)

TP270.A1D47 1987 660.2′804 87-14291
ISBN 0-8412-1405-0

Copyright © 1987

American Chemical Society

All Rights Reserved. The appearance of the code at the bottom of the first page of each chapter in this volume indicates the copyright owner's consent that reprographic copies of the chapter may be made for personal or internal use or for the personal or internal use of specific clients. This consent is given on the condition, however, that the copier pay the stated per copy fee through the Copyright Clearance Center, Inc., 27 Congress Street, Salem, MA 01970, for copying beyond that permitted by Sections 107 or 108 of the U.S. Copyright Law. This consent does not extend to copying or transmission by any means—graphic or electronic—for any other purpose, such as for general distribution, for advertising or promotional purposes, for creating a new collective work, for resale, or for information storage and retrieval systems. The copying fee for each chapter is indicated in the code at the bottom of the first page of the chapter.

The citation of trade names and/or names of manufacturers in this publication is not to be construed as an endorsement or as approval by ACS of the commercial products or services referenced herein; nor should the mere reference herein to any drawing, specification, chemical process, or other data be regarded as a license or as a conveyance of any right or permission, to the holder, reader, or any other person or corporation, to manufacture, reproduce, use, or sell any patented invention or copyrighted work that may in any way be related thereto. Registered names, trademarks, etc., used in this publication, even without specific indication thereof, are not to be considered unprotected by law.

PRINTED IN THE UNITED STATES OF AMERICA

ACS Symposium Series

M. Joan Comstock, *Series Editor*

1987 Advisory Board

Harvey W. Blanch
University of California—Berkeley

Alan Elzerman
Clemson University

John W. Finley
Nabisco Brands, Inc.

Marye Anne Fox
The University of Texas—Austin

Martin L. Gorbaty
Exxon Research and Engineering Co.

Roland F. Hirsch
U.S. Department of Energy

G. Wayne Ivie
USDA, Agricultural Research Service

Rudolph J. Marcus
Consultant, Computers & Chemistry Research

Vincent D. McGinniss
Battelle Columbus Laboratories

W. H. Norton
J. T. Baker Chemical Company

James C. Randall
Exxon Chemical Company

E. Reichmanis
AT&T Bell Laboratories

C. M. Roland
U.S. Naval Research Laboratory

W. D. Shults
Oak Ridge National Laboratory

Geoffrey K. Smith
Rohm & Haas Co.

Douglas B. Walters
National Institute of Environmental Health

Foreword

The ACS SYMPOSIUM SERIES was founded in 1974 to provide a medium for publishing symposia quickly in book form. The format of the Series parallels that of the continuing ADVANCES IN CHEMISTRY SERIES except that, in order to save time, the papers are not typeset but are reproduced as they are submitted by the authors in camera-ready form. Papers are reviewed under the supervision of the Editors with the assistance of the Series Advisory Board and are selected to maintain the integrity of the symposia; however, verbatim reproductions of previously published papers are not accepted. Both reviews and reports of research are acceptable, because symposia may embrace both types of presentation.

Contents

Preface

THIS BOOK DESCRIBES the assessment of the combined hazards of toxic chemical and explosives facilities. The principal considerations regarding explosive and toxic chemical outputs are blast pressure, fragmentation, thermal parameters, and toxic chemical exposures. The book provides design considerations for protecting workers from these outputs and for protecting property within and away from the facilities. Practical examples and protection principles from multiple disciplines are given; these deal with practices, training, site selection, quantity-distance separation, downwind hazard-prediction models, storage methods, and disposal. In addition, methods of measuring and controlling the exposure of workers to toxic chemicals and the development and implementation of engineering and construction features are addressed.

RALPH A. SCOTT, JR.
Defense Explosives Safety Board
Alexandria, VA 22331

LAURENCE J. DOEMENY
National Institute for Occupational Safety and Health
Cincinnati, OH 45226

May 13, 1987

BLAST PRESSURE AND FRAGMENTATION EFFECTS

Chapter 1

Blast Pressure Effects: An Overview

W. E. Baker

Wilfred Baker, Inc., P.O. Box 6477, San Antonio, TX 78209

This keynote paper gives a general discussion of blast waves developed by high explosive detonations, their effects on structures and people, and risk assessment methods. The properties of free-field waves and normally and obliquely reflected waves are reviewed. Diffraction around block shapes and slender obstacles is covered next. Blast and gas pressures from explosions within vented structures are summarized.

Simplified methods of estimating damage to structures by blast waves appear next, followed by methods of estimating blast spalling for strong blasts.

Prediction curves or graphs are given for external blast wave properties, and internal blast and gas transient pressures.

Practical techniques for explosion containment and venting are discussed, and the topic of risk assessment for explosives facilities is reviewed.

A selected reference list closes the paper.

Blast Pressures

Basics of Free-Field Blast Waves. The most severe types of energy releases which can occur in toxic chemical and explosives facilities are explosions of high explosive materials. When such materials are initiated by some stimulus, they may burn, deflagrate or detonate. Detonation is by far the most severe of these three chemical reactions, so it is usually assumed to occur in accident situations, unless one can prove otherwise quite conclusively.

A detonation wave is a very rapid wave of chemical reaction which, once it is initiated, travels at a stable supersonic speed, called the detonation velocity, in a high explosive. Typically, detonation velocities for pressed or cast high explosives range from

0097-6156/87/0345-0002$14.40/0
© 1987 American Chemical Society

22,000 - 28,000 ft/sec. As the detonation wave progresses through the condensed explosive, it converts the explosive within a fraction of a microsecond into very hot, dense, high pressure gas. Pressures immediately behind the detonation front range from 2,700,000 - 4,900,000 psi.(These pressures are called Chapman-Jouguet, or CJ, pressures.)

The most important single parameter for determining air blast wave characteristics of high explosives is the total heat of detonation, E. This quantity is, in general, directly proportional to the total weight W or mass M of the explosive. Any given explosive has a specific heat of detonation, ΔH_e per unit weight or mass, which can be either calculated from chemical reaction formulas or measured calorimetrically (see References 1-3). So E equals $W \cdot \Delta H_e$ or $M \cdot \Delta H_e$, depending on units for ΔH_e. Values for ΔH_e for many explosives are given in References 1 and 4.

If the detonating explosive is bare, the detonation wave propagates out into the surrounding air as an intense shock or blast wave, and is driven by the expanding hot gases which had been the explosive material. If it is encased, the detonation wave simply overpowers the casing material, and drives it outward at high velocity until the casing fragments. The high pressure gases then vent out past the casing fragments and again drive a strong blast wave into the surrounding atmosphere.

As the blast wave expands, it decays in strength, lengthens in duration, and slows down, both because of spherical divergence and because the chemical reaction is over, except for afterburning as the hot explosion products mix with the surrounding air.

Good descriptions of the characteristics of air blast waves appear in References 5-7. The description here is paraphrased from Reference 5.

As a blast wave passes through the air or interacts with and loads a structure or target, rapid variations in pressure, density, temperature and particle velocity occur. The properties of blast waves which are usually defined are related both to the properties which can be easily measured or observed and to properties which can be correlated with blast damage patterns. It is relatively easy to measure shock front arrival times and velocities and entire time histories of overpressures. Measurement of density variations and time histories of particle velocity are more difficult, and few reliable measurements of temperature variations exist.

Classically, the properties which are usually defined and measured are those of the undisturbed or side-on wave as it propagates through the air. Figure 1 shows graphically some of these properties in an ideal wave. Prior to shock front arrival, the pressure is ambient pressure p_o. At arrival time t_a, the pressure rises quite abruptly (discontinuously, in an ideal wave) to a peak value $P_s + p_o$. The pressure then decays to ambient in total time $t_a + t_d$, drops to a partial vacuum and eventually returns to p_o. The quantity P_s is usually termed the peak side-on overpressure, or merely the peak overpressure. The portion of the time history above initial ambient pressure is called the positive phase, of duration t_d. That portion below p_o is called the negative phase. Positive specific impulse, defined by

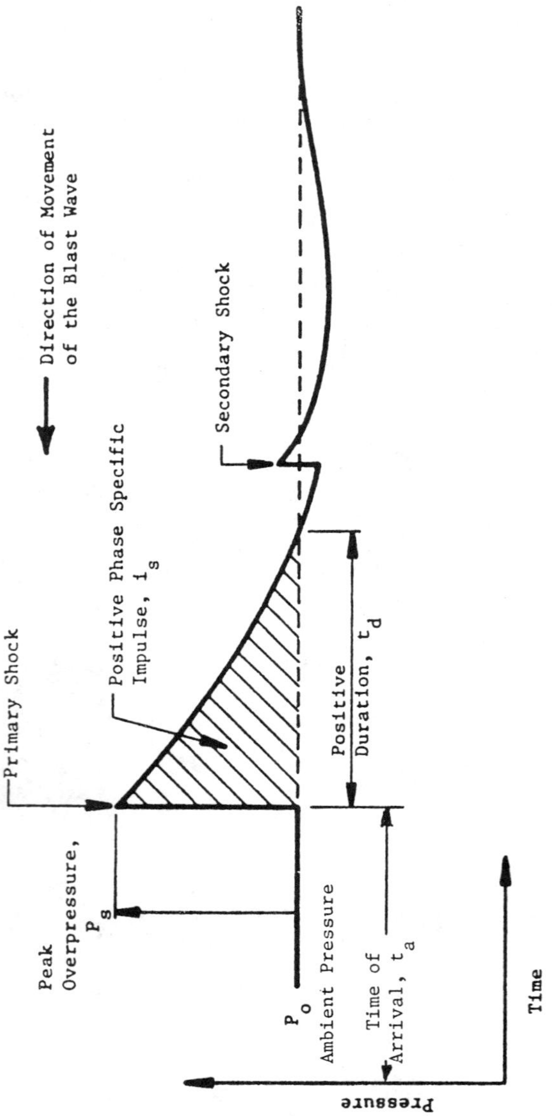

Figure 1. Idealized Profile of a Blast Wave from a Condensed High Explosive.(Courtesy Oyez Scientific and Technical Services Ltd.)

$$i_s = \int_{t_a}^{t_a + t_d} [p(t) - p_o]dt \qquad (1)$$

is also a significant blast wave parameter. This impulse is shown by the cross-hatched area in Figure 1. (The units of i_s are force times time divided by length squared, or pressure times time. They are, therefore specific impulse or impulse per unit area, rather than true impulse, which has units of force times time.)

In most blast studies, the negative phase of the blast wave does not affect damage and is ignored, and only blast parameters associated with the positive phase are considered or reported. The ideal side-on parameters almost never represent the actual pressure loading applied to structures or targets following an explosion. So a number of other properties are defined to either more closely approximate real blast loads or to provide upper limits for such loads. (The processes of reflection and diffraction will be discussed later.) Properties of free-field blast waves other than side-on pressure which can be important in structural loading are:

Density, ρ
Particle velocity, u
Shock front velocity, U
Dynamic pressure $q = \rho u^2/2$

Because of the importance of the dynamic pressure q in drag or wind effects and target tumbling, it is often reported as a blast wave property. In some instances drag specific impulse i, defined as

$$i_d = \int_{t_a}^{t_a + t_d} q\ dt = \frac{1}{2} \int_{t_a}^{t_a + t_d} \rho u^2\ dt \qquad (2)$$

is also reported.

Although it is possible to define the potential or kinetic energy in blast waves, it is not customary in air blast technology to report or compute these properties. For underwater explosions, the use of "energy flux density" is more common. This quantity is given approximately by

$$E_f = \frac{1}{\rho_o a_o} \int_{t_a}^{t_a + t_d} [p(t) - p_o]^2\ dt \qquad (3)$$

where ρ_o and a_o are density and sound velocity in water ahead of the shock.

At the shock front in free air, a number of wave properties are interrelated through the Rankine-Hugoniot equations. These three equations are (Reference 5):

$$\rho_s(u_s-U) = \rho_o(u_o-U) \tag{4}$$

$$\rho_s(u_s-U)^2 + p_s = \rho_o(u_o-U)^2 + p_o \tag{5}$$

$$\left(\frac{1}{2}u_o^2 + e_o\right)\left(u_o-U\right) + p_o u_o = \left(\frac{1}{2}u_s^2 + e_s\right)\left(u_s - U\right) + p_s u_s \tag{6}$$

In these equations, subscript s refers to peak quantities immediately behind the ideal shock front, e is internal energy, and

$$p_s = P_s + p_o \tag{7}$$

Scaling of the properties of blast waves from explosive sources is a common practice, and anyone who has even a rudimentary knowledge of blast technology utilizes these laws to predict the properties of blast waves from large-scale explosions based on tests on a much smaller scale. Similarly, results of tests conducted at sea level ambient atmospheric conditions are routinely used to predict the properties of blast waves from explosions detonated under high altitude conditions.

The most common form of blast scaling is Hopkinson-Cranz or "cube-root" scaling. This law, first formulated by B. Hopkinson (Reference 8) and independently by C. Cranz (Reference 9), states that self-similar blast waves are produced at identical scaled distances when two explosive charges of similar geometry and of the same explosive, but of different sizes, are detonated in the same atmosphere. It is customary to use as a scaled distance a dimensional parameter,

$$Z = R/E^{1/3} \tag{8}$$

or

$$Z = R/W^{1/3} \tag{9}$$

where R is the distance from the center of the explosive source, E is the total heat of detonation of the explosive and W is the total weight of a standard explosive such as TNT. The correct equation, Equation 8 or 9, will be apparent in the problem. Figure 2 shows schematically the implications of Hopkinson-Cranz blast wave scaling. An observer located at a distance R from the center of an explosive source of characteristic dimension d will be subjected to a blast wave with amplitude P, duration t_d, and a characteristic time history. The integral of the pressure-time history is the impulse i. The Hopkinson-Cranz scaling law then states that an observer stationed at a distance λR from the center of a similar explosive source of characteristic dimension λd detonated in the same atmosphere will feel a blast wave of "similar" form with amplitude P, duration λt_d and impulse λi. All characteristic times are scaled by the same factor as the length scale factor λ. In Hopkinson-Cranz scaling, pressures, temperatures, densities and velocities are unchanged at homologous times. This scaling law has been thoroughly verified by many experiments conducted over a large range of explosive charge energies. A much more complete discussion of this law and demonstration of its applicability is given in Chapter 3 of Reference 5.

The blast scaling law which is almost universally used to predict characteristics of blast waves from explosions at high altitude is that of Sachs (Reference 10). Sachs' law states that dimensionless overpressure and dimensionless impulse can be expressed as unique functions of a dimensionless scaled distance, where the dimensionless parameters include quantities which define the ambient atmospheric conditions prior to the explosion. Sachs' scaled pressure is

$$\overline{P} = (P/p_o) \qquad (10)$$

Sachs' scaled impulse is defined as

$$\overline{i} = \frac{i a_o}{E^{1/3}\, p_o^{2/3}} \qquad (11)$$

where a_o is ambient sound velocity. These quantities are a function of dimensionless scaled distance, defined as

$$\overline{R} = R\left(\frac{p_o}{E}\right)^{1/3} \qquad (12)$$

Both scaling laws apply to reflected blast wave parameters, as well as side-on parameters. (Note that, if charge weight W is used instead of energy E, these parameters have dimensions.)

Basics of Reflection and Diffraction Processes

Normal Reflection. An upper limit to blast loads is obtained if one interposes an infinite, rigid wall in front of the wave, and reflects the wave normally. All flow behind the wave is stopped, and pressures are considerably greater than side-on. The pressure in normally reflected waves is usually designated $p_r(t)$, and the peak reflected overpressure, P_r. The integral of overpressure over the positive phase, defined in Equation (13), is the reflected specific impulse i_r. Durations of the positive phase of normally reflected waves are almost the same as for side-on waves, t_d. The parameter i_r has been measured closer to high explosive blast sources than have most blast parameters.

$$i_r = \int_{t_a}^{t_a + t_d} [p_r(t) - p_o]\, dt \qquad (13)$$

The Hopkinson-Cranz scaling law described earlier applies to scaling of reflected blast wave parameters just as well as it does to side-on waves. That is, all reflected blast data taken under the same atmospheric conditions for the same type of explosive source can be reduced to a common base for comparison and prediction. Sachs' law for reflected waves fails close to high explosive blast sources but it does apply beyond about ten charge radii.

For shock waves weak enough that air behaves as a perfect gas, there is a fixed and well-known relation between peak reflected overpressure and peak side-on overpressure (References 5 and 11).

$$\overline{P}_r = 2\,\overline{P}_s + \frac{(\gamma+1)\,\overline{P}_s^{\,2}}{(\gamma-1)\,\overline{P}_s+2} \qquad (14)$$

$$\overline{P}_s = P_s/p_o \qquad (15)$$

$$\overline{P}_r = P_r/p_o \qquad (16)$$

At low incident overpressures ($P_s \rightarrow 0$), the reflected overpressure approaches the acoustic limit of twice the incident overpressure. If one were to assume a constant $\gamma = 1.4$ for air for strong shocks, the upper limit would appear to be $P_r = 8P_s$. But, air ionizes and dissociates as shock strengths increase, and γ is not constant. In fact, the real upper limit ratio is not exactly known, but is predicted by Doering and Burkhardt (Reference 11) to be as high as 20. Brode (Reference 12) has also calculated this ratio for normal reflection of shocks in sea level air, assuming air dissociation and ionization.

A curve plotted from an equation in Ref. 12 is reproduced here as Figure 3. Above P_s = 100 psi and standard atmospheric conditions, Eq. (14) is increasingly in error compared to this curve, and should not be used. (Note that, at the surface of a spherical TNT charge at sea level, Ref. 12 and Figure 3 give P_r/P_s = 13.92.)

Oblique Reflection. Although normally incident blast wave properties usually provide upper limits to blast loads on structures, the more usual case of loading of large, flat surfaces is represented by waves which strike at oblique incidence. Also, as a blast wave from a source some distance from the ground reflects from the ground, the angle of incidence must change from normal to oblique as the shock moves across the ground surface.

Oblique reflection is classed as either regular or Mach reflection, dependent on incident angle and shock strength. Geometries of these two cases are shown in Figures 4 and 5 from Reference 13. In regular reflection, the incident shock travels into still air (Region One) at velocity U, with its front making the angle of incidence α_I with respect to the wall. Properties behind this front (Region Two) are those for a free air shock. On contact with the wall, the flow behind the incident shock is turned, because the component normal to the wall must be zero, and the shock is reflected from the wall at a reflection angle α_R that is different from α_I. Conditions in Region Three indicate reflected shock properties. A pressure transducer flush-mounted in the wall would record only the ambient and reflected wave pressures (direct jump from Region One to Region Three) as the wave pattern traveled along the wall; whereas, one mounted at a short distance from the wall would record the ambient pressure, then the incident wave pressure, and finally the reflected wave pressure.

There is some critical angle of incidence, $\alpha_{extreme}$ dependent on shock strength, above which regular reflection cannot occur. In 1877, Ernst Mach showed that the incident and reflected shocks would coalesce to form a third shock. Because of the geometry of the shock fronts, they were termed the Mach V or Mach Y, with the single shock formed by the coalesced incident and reflected shocks normally called the Mach stem. The geometry of Mach reflection is shown in Figure 5. In addition to the incident and reflected shocks I and R, we now have the Mach shock M; the junction T of the three shocks is called the triple point. In addition, there is also a slipstream S, a boundary between regions of different particle velocity and different density, but the same pressure. When α_I in Figure 4 exceeds $\alpha_{extreme}$, the Mach wave M is formed at the wall and grows as the shock systems move along the wall with the locus of the triple point being a straight line AB.

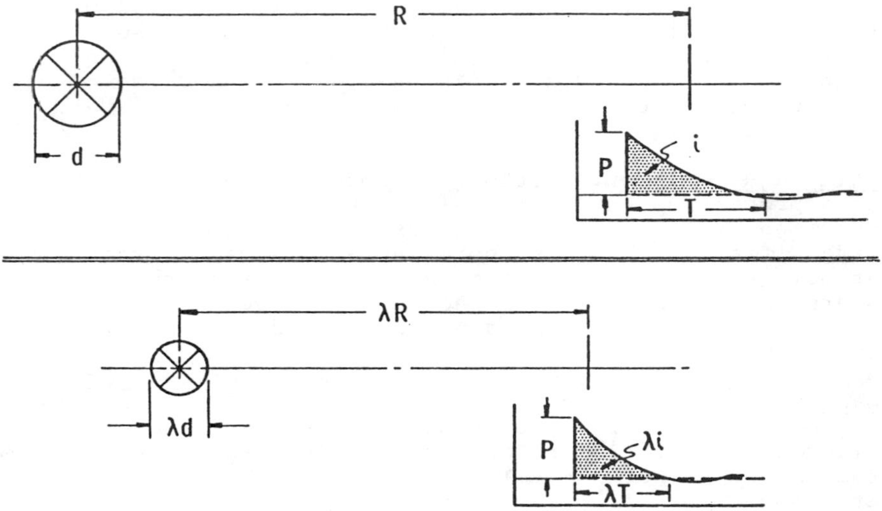

Figure 2. Hopkinson-Cranz Blast Wave Scaling.

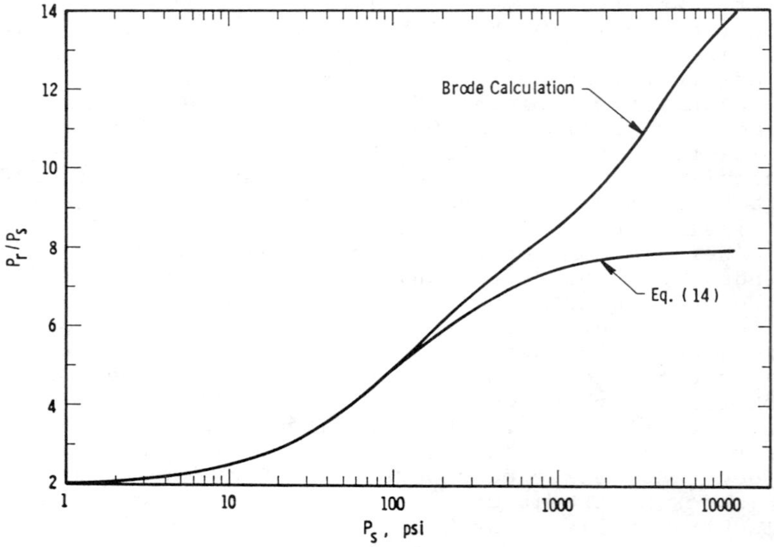

Figure 3. Ratio of P_r/P_s Versus P_s for Sea Level Ambient Pressure.

Harlow and Amsden (Reference 14) present a resume of theory and experiment on regular reflection and the limit of regular reflection (which is also the start of Mach reflection). A useful curve from their paper is given here. Figure 6 gives angle of reflection α_R as a function of angle of incidence α_I in the regular regime. The parameter ξ is defined as

$$\xi = \frac{p_o}{p_s + p_o} \qquad (17)$$

[Harlow and Amsden (Ref. 14) call ξ the shock strength, but it is, in fact, the <u>inverse</u> of shock strength.] Inverting Equation (17) we also have the relation

$$\overline{P}_s = \frac{p_s}{p_o} = \frac{1}{\xi} - 1 \qquad (18)$$

<u>Diffraction</u>. When a blast wave encounters a finite obstacle, it is partially reflected but also <u>diffracts</u> around the obstacle. This process is described here.

The process of diffraction of a blast wave around a rectangular block object, such as a simple building shape, is well described in Ref. 7, and is paraphrased here.

When the front of an air blast wave strikes the face of a structure reflection occurs. As a result the overpressure builds up rapidly to at least twice (and generally several times) that in the incident wave front. The actual pressure attained is determined by various factors, such as the peak overpressure of the incident blast wave and the angle between the direction of motion of the wave and the face of the structure. The pressure increase is due to the conversion of the kinetic energy of the air behind the shock front into internal energy as the rapidly moving air behind the shock front is decelerated at the face of the structure. The high pressure region expands outward towards the surrounding regions of lower pressure.

As the wave front moves forward, the reflected overpressure on the face of the structure drops rapidly to the side-on overpressure, plus an added drag force due to the wind (dynamic) pressure. At the same time, the air pressure wave bends or "diffracts" around the structure, so that the structure is eventually engulfed by the blast, and approximately the same pressure is exerted on the sides and the roof. The front face, however, is still subjected to wind pressure, although the back face is shielded from it.

The developments described above are illustrated in a simplified form in Figs. 7a, b, c, d, e; this shows, in plan, successive stages of a structure without openings which is being struck by an air blast wave moving in a horizontal direction. In Fig. 7a the wave front is seen approaching the structure with the direction of motion perpendicular to the face of the structure exposed to the blast. In Fig. 7b the wave has just reached the front face, producing a high

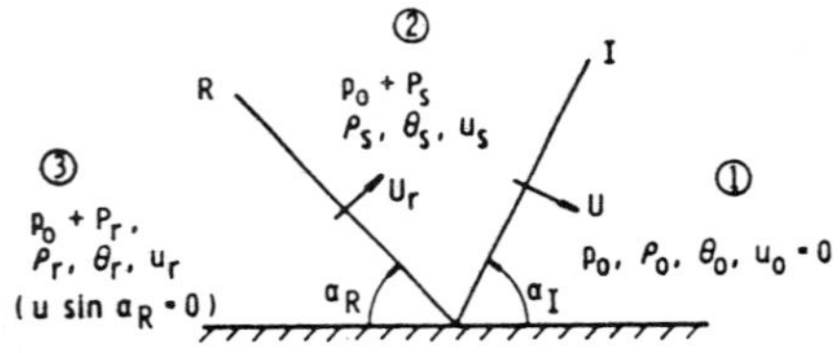

Figure 4. Regular Oblique Reflection of a Plane Shock from a Rigid Wall.(Reference 13)

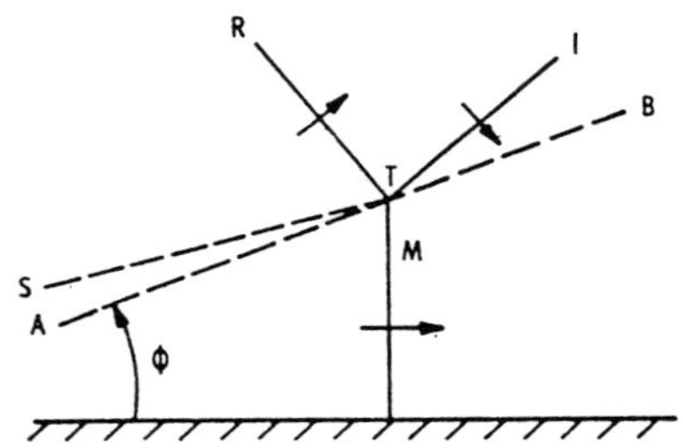

Figure 5. Mach Reflections From a Rigid Wall. (Reference 13)

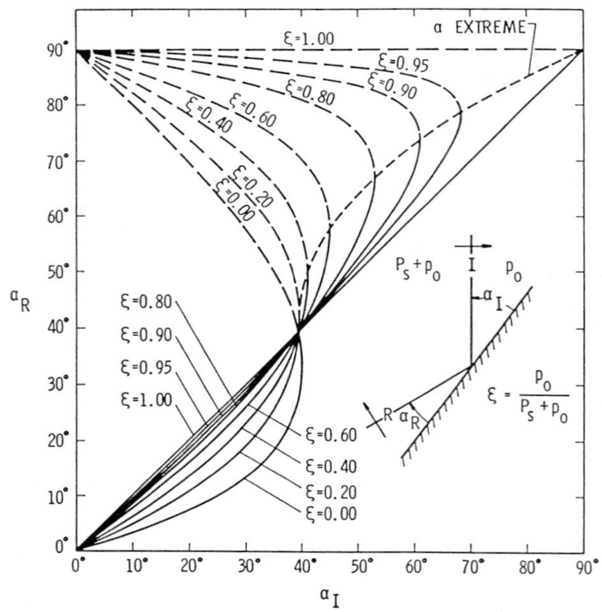

Figure 6. Angle of Incidence Versus Angle of Reflection for Shocks of Different Strengths Undergoing Regular Reflection. (Reference 14)

reflected overpressure. In Fig. 7c the blast wave has proceeded about halfway along the structure. In Fig. 7d the wave front has just passed the rear of the structure. The pressure on the front face has dropped to some extent while the pressure is building up on the back face as the blast wave diffracts around the structure. Finally, when the wave front has passed completely, as in Fig. 7e, approximately equal air pressures are exerted on the sides and top of the structure. A pressure difference between front and back faces, due to the wind forces, will persist, however, during the whole positive phase of the blast wave (Fig. 7f). If the structure is oriented at an angle to the blast wave, the pressure would immediately be exerted on two faces, instead of one, but the general characteristics of the blast loading would be similar to that just described (Figs. 7g, h, and i).

The pressure differential between the front and back faces will have its maximum value when the blast wave has not yet completely surrounded the structure, as in Figs. 7c, and d or g and h. Such a pressure differential will produce a lateral (or translational) force tending to cause the structure to deflect and thus move bodily, usually in the same direction as the blast wave. This force is known as the "diffraction loading" because it operates while the blast wave is being diffracted around the structure.

When the blast wave has engulfed the structure (Fig. 7e or 7i), the pressure differential is small and the loading is due almost entirely to the drag pressure exerted on the front face. The actual pressures on all faces of the structure are in excess of the ambient atmospheric pressure and will remain so, although decreasing steadily, until the positive phase of the blast wave has ended. Hence, the diffraction loading on a structure without openings is eventually replaced by an inwardly directed pressure, i.e., a compression or squeezing action, combined with the dynamic pressure of the blast wave. In a structure with no openings, the loading will cease only when the overpressure drops to zero.

For blast waves from relatively small explosion sources, the diffraction phase of the loading may dominate, and the drag phase may be relatively or entirely unimportant, because the diffraction times may be as long as or greater than drag pressure durations.

Reference 7 gives explicit procedures for calculating diffracted loads on surfaces of box-shaped structures, and they will not be repeated here. But, we do reproduce several formulas for diffraction times from this reference. These are

$$t_1 = \frac{4\ S}{(1+R)a_o} \tag{19}$$

$$t_2 = \frac{L}{U} \tag{20}$$

$$t_3 = \frac{L}{U} + \frac{2\,S}{a_o} \tag{21}$$

where S is the lesser of H or B/2 in Figure 8, G is the greater of H or B/2, R is S/G, L is block length, and U is shock front velocity.

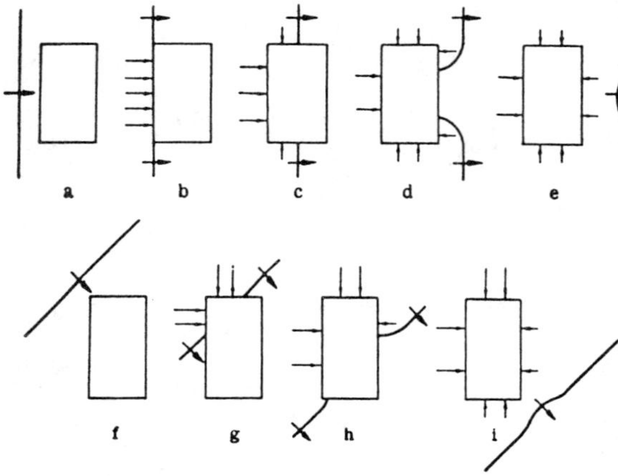

Figure 7. Stages in the Diffraction of a Blast Wave by a Structure without Openings. (Plan View) (Ref. 7)

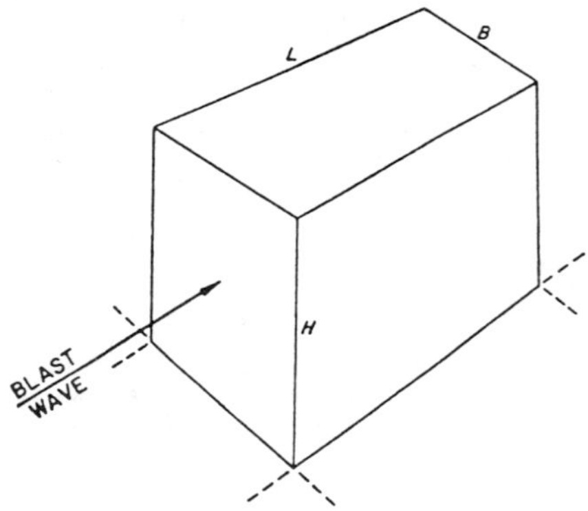

Figure 8. Representation of a Closed Box-Like Structure.

If the structure being loaded by the blast wave is a slender member or object such as a column, I-beam, or stack, then the diffraction times indicated by Equations (19) - (21) give short times because transverse dimensions are small. The diffraction around such objects is illustrated in Figure 9, with stages similar to those described for diffraction around a block structure. Here, the diffraction phase is almost always shorter than the drag phase, and we are interested primarily in the net transverse pressure loading on the slender structure or object. A simplified time history of this loading appears in Figure 10. Methods for calculating this net pressure loading are given in Ref. 15, for TNT blast sources.

Gas Pressures in Vented and Unvented Enclosures

A recent review on the topic of the relatively long-term gas pressures which develop for explosions within enclosures appears in Ref. 16. That material is summarized here.

For explosions in enclosures involving high explosives, solid propellants, high explosive with combustible materials in contact, or combustible mist, dust, or gaseous explosive mixtures, the long-duration gas pressures caused by confinement of the products of the explosions can be the dominant loads causing structural failure. These quasi-static pressures are determined by the total heat energy in the explosive and/or combustible source, the volume of the enclosure, the vent area and the vent panel configuration, the mass per unit area of vent covers, and the initial ambient conditions within the enclosure.

Here, we concentrate on the gas pressures developed for high explosive detonations within vented and unvented enclosures, and these explosives plus nearby combustible materials. There is a voluminous literature on pressures and the effects of venting for confined explosions with only combustible gases and dusts in air, but that topic seems outside the scope of this book, and is not discussed here.

The loading from an explosive charge detonated within a structure consists of two phases. The initial phase consists of several high amplitude, short duration, reflected pressure shocks. This phase of the loading is very geometry dependent, with the highest loads generally occuring on the surfaces nearest the charge. On each reflection, the shock strength is attenuated until at some point the internal pressure has settled to a slowly decaying level. This is the quasi-static pressure loading phase. This phase is characterized by essentially uniform pressures throughout the structure at any point in time. The rate of quasi-static pressure decay is a function of the vent area, structure volume and the nature of the explosive source (e.g., propellant versus explosive).

A typical pressure trace obtained during an internal explosion in a vented structure is shown in Figure 11. Traditionally (Ref. 17), the peak quasi-static pressure is established by fitting a smooth line through the data beginning at the end of the pressure trace and extending back towards time zero, the time of charge ignition. This line is shown in Figure 11 as a solid line. The peak P_{qs} is then taken as the intersection of the fitted line and a vertical line at time zero (shown as a dotted line in the figure). This point is labeled A in Figure 11. For a vented structure, a

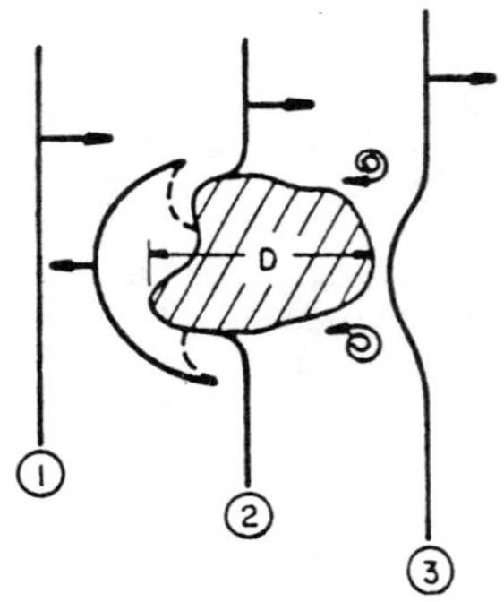

Figure 9. Interaction of Blast Wave with Slender Object.

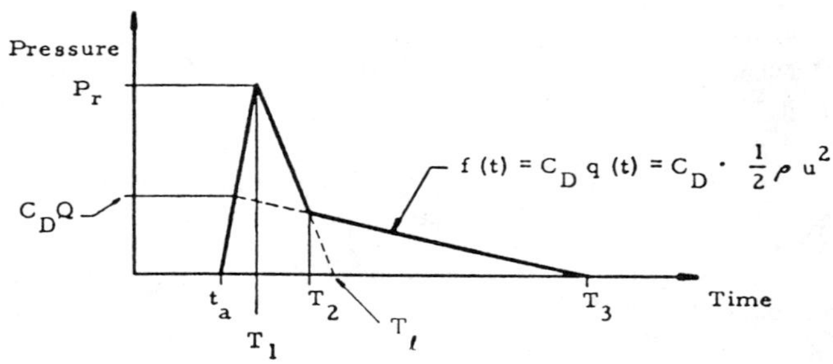

Figure 10. Time History of Net Transverse Pressure on Object during Passage of a Blast Wave.

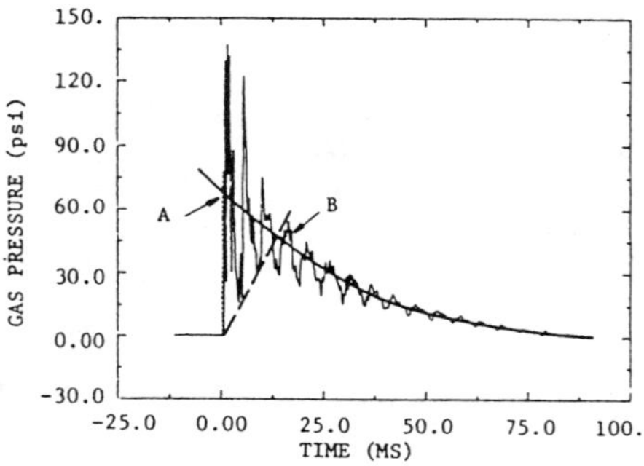

Figure 11. Typical Pressure Record from an Internal Explosion in a Vented Structure.

more appropriate technique has been suggested (Ref. 13 and 18). This method is applied by drawing a ramp increase in pressure extending from time zero, which follows the base of the pressure shocks. This line is shown as a dashed line in Figure 11. The intersection of the ramp pressure increase with the line fitted through the pressure decay is the peak quasi-static pressure. This point is labeled B in the figure. For explosions inside sealed enclosures, points A and B will have nearly the same ordinates, whereas for explosions with increasing vent areas, the difference in ordinates between points A and B increases.

In Ref. 18, a very complete analysis of gas pressures from internal explosion data was presented. The authors performed a similitude analysis to determine the functional form of the quasi-static pressure, as a function of the physical parameters pertaining to the problem of an internal explosion inside a vented structure.

This analysis gave the following dimensionless functional forms:

$$\bar{p} = \frac{p_{QS}+p_o}{p_o} \qquad (22)$$

$$\bar{p} = f\left[\frac{W}{p_o V}\right] \qquad (23)$$

$$\bar{T} = \left(\frac{ta_o}{V^{1/3}}\right)\left(\frac{\alpha_{eff}A}{V^{2/3}}\right) = g\left[\frac{p}{p_o}\,,\ 1\right] \qquad (24)$$

$$\bar{i}_g = \frac{i_g a_c \alpha_{eff} A}{p_o V} = h\left[\frac{p}{p_o}\right] \qquad (25)$$

In these expressions,

p = absolute peak gas pressure
P_{QS} = gage peak gas pressure
p_o = atmospheric pressure
W = charge total energy (<u>not</u> weight)
V = enclosure volume
$\alpha_{eff}A$ = effective vent area
t = venting time
a_o = sound speed
i_g = gas impulse
f,g,h = functional forms

The authors of Ref. 18 fitted data from over 175 experiments to the scaled vented pressure parameters, using total heats of explosion for W. Graphs from that paper will be shown later.

Most gas pressure parameters for vented HE explosions apply for open vents and the special venting configurations developed for suppressive shields (Refs. 17 and 19). If vents are covered with blowout or frangible covers, the peak gas pressures are essentially the same as in unvented structures, but venting times and gas impulses can be altered (increased), depending on the vent area, mass per unit area of the vent cover, and initial shock reflected impulse loading of the vent cover. The staff of the Naval Civil Engineering Laboratory has conducted a number of analytic and experimental studies to determine these effects, and has developed methods for predicting the resulting gas pressure loads. The methods will be discussed later. Detailed graphs are too numerous to include. They will appear in the revision to TM5-1300.

Combustion of gas-air mixtures within enclosures has long been known to produce significant pressure increases because of air heating by all or part of the heat of combustion of the gaseous fuel. So, it should not be surprising that combustibles near or in intimate contact with high explosives detonated in enclosures can in many instances raise the gas pressures well above the gas pressures from detonations of only the high explosives. But, it is surprising that there has been little testing to measure and allow prediction for such increase. One of the few such test programs is reported in Ref. 20, with some results summarized in Figure 12. The effect has been observed for a variety of combustible materials, but no variations in charge to combustible mass, charge type, structure volume, or degree of venting have been tested. The implications of the data accumulated so far are that quasi-static loading calculations should include estimates of contributions from the burning of combustible materials whenever such materials are expected to be in intimate contact with HE sources.

Damage Mechanisms

The P-i Curve Concept and Applications. We hope that Section I of this chapter demonstrates the Dynamic and transient nature of the blast waves caused by explosives detonations, and the resulting pressure loads they can apply to various structures or objects. Because these loads are usually suddenly applied, and because they last from fractions of a millisecond to at most seconds, the response of or damage to loaded structures or objects is almost always dynamic. So, usually structural response or damage is dependent not only on the amplitude (peak overpressure) of the applied blast loading, the loaded area and the structural strength; but also on the mass or inertia of the structure, and either the duration of the transient pressure loading or the applied specific impulse.

These concepts are probably most simply developed by first calculating the response of very simple dynamic mechanical systems. This has been done in Refs. 15 and 21, and the reader is referred to either of these references for detailed development.

Consider the simple elastic system of Fig. 13. Equations of motion under the applied (simplified) force pulse can be easily written and solved (see Refs. 15 and 21), and a dimensionless form of the maximum response $\overline{X}_{max}$ can be plotted versus another dimensionless ratio which relates loading time T to structural natural period (Figure 14). In these two figures, the various symbols represent:

P^* = peak applied force (not pressure)
t = time
T = effective blast wave duration
m = mass

k = spring constant
x = displacement
ω = circular vibration frequency
τ = vibration period

In Figure 14, we see that the scaled maximum response reaches asymptotic relations for both small and large time ratios.

The same solution presented in Fig. 14 can easily be recast (see Ref. 22) into another form, as in Fig. 15. Note here that the maximum scaled response curve is now essentially a rectangular hyperbola with one asymptote which depends only on the level of applied peak force and another asymptote which depends only on the level of applied total impulse. In the intermediate loading regime (the "knee" of the hyperbola), response determination requires knowledge of both peak force and total impulse.

This P*-I type of response curve can also be easily shown to apply to a simple rigid-plastic mechanical system, in the manner shown in Figure 16 (see Refs. 15 and 22). Here, the spring in the system is replaced with a pure Coulomb friction element, with resisting force f, which is independent of displacement once the mass starts to move. All other symbols are defined above.

Although the curves in Figures 13-15 were developed for transient loads defined by total applied forces and impulses, we could as easily have developed them by initially specifying an applied pressure transient loading, with its accompanying specific impulse, plus a loaded area. So, the concept certainly applies to simple structures under blast loading. The important inferences to be drawn from the simple analyses are that structures respond primarily to peak overpressure if their vibration periods are much shorter than the blast duration, while they respond primarily to specific impulse if their vibration periods are much longer than the blast duration. If these two times are about equal, then both blast loading quantities are important.

Biggs (Ref. 21) discusses responses of simple dynamic systems in great detail, including the important intermediate case of elastic, perfectly-plastic systems. He also presents dimensionless response curves for various levels of elastic-plastic response, and for several different regular pulse shapes.

Does this concept of a P-i diagram as a measure of response or damage work for complex structures, as well as simple ones? Indeed it does, as can be shown by the fits made in Britain for bomb damage to houses, following World War II. These fits, illustrated in Fig. 17, now form part of the basis for the British Quantity-Distance tables for explosives safety.

If one can calculate or measure an "isodamage curve" for a structure or structural element, i.e., an hyperbola similar to Figure 17, one can plot it as an overlay on those combinations of peak overpressure and specific impulse which result from detonating various explosive charge masses or energies at various distances, and graphically convert the isodamage curve to a set of combinations of charge masses and distances which cause this damage. Figure 18 is an example for a light structure which is susceptible to damage from small mass charges. Some specific examples used to calculate

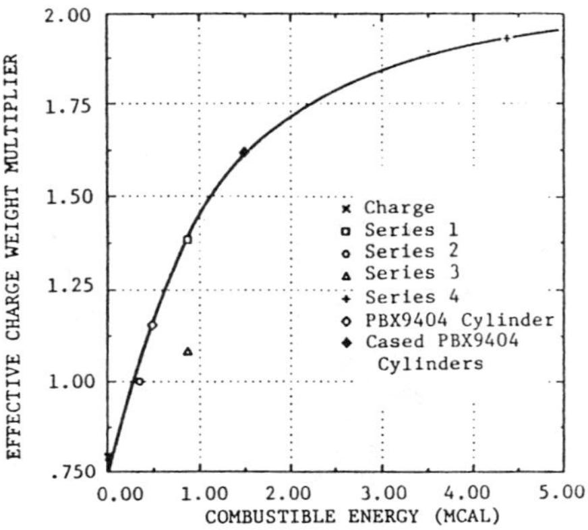

Figure 12. Effective Charge Weight Multiplier for Combustibles. (Ref. 19)

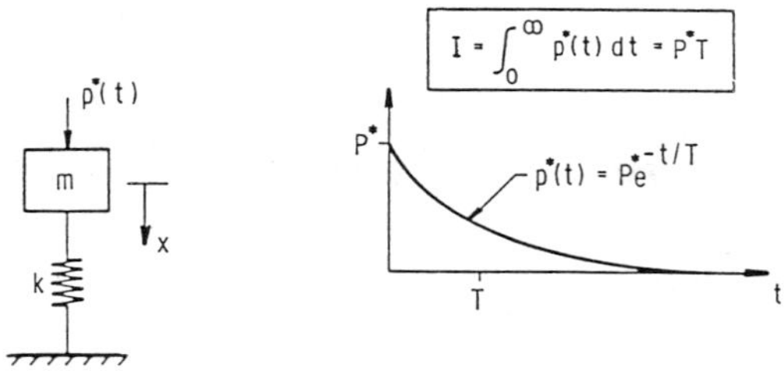

Figure 13. Linear Oscillator Loaded by a Blast Wave.

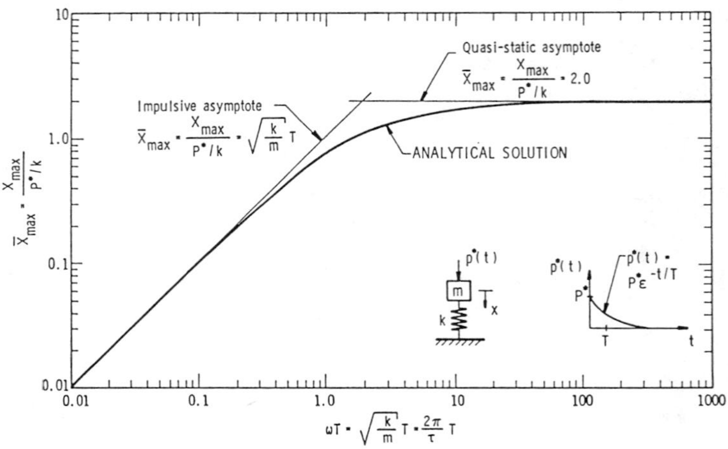

Figure 14. Shock Response for Blast-Loaded Elastic Oscillator.

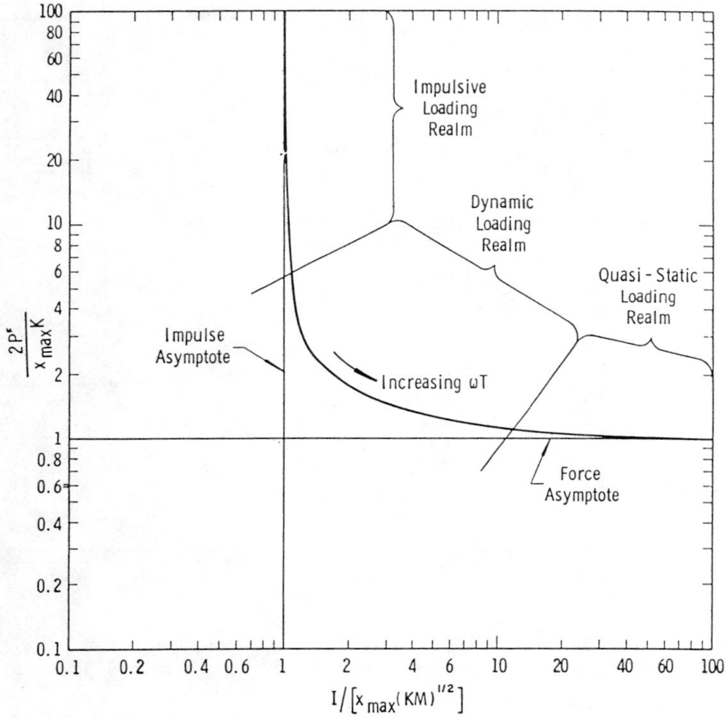

Figure 15. P*-I Diagram for Blast-Loaded Elastic Oscillator.

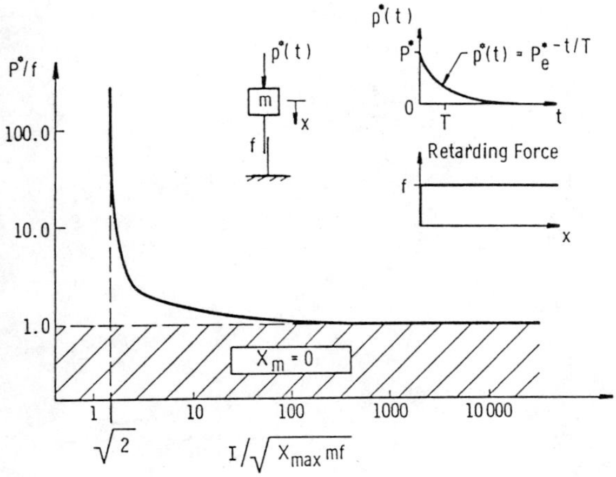

Figure 16. P*-I Diagram for Blast Loaded Rigid-Plastic System.

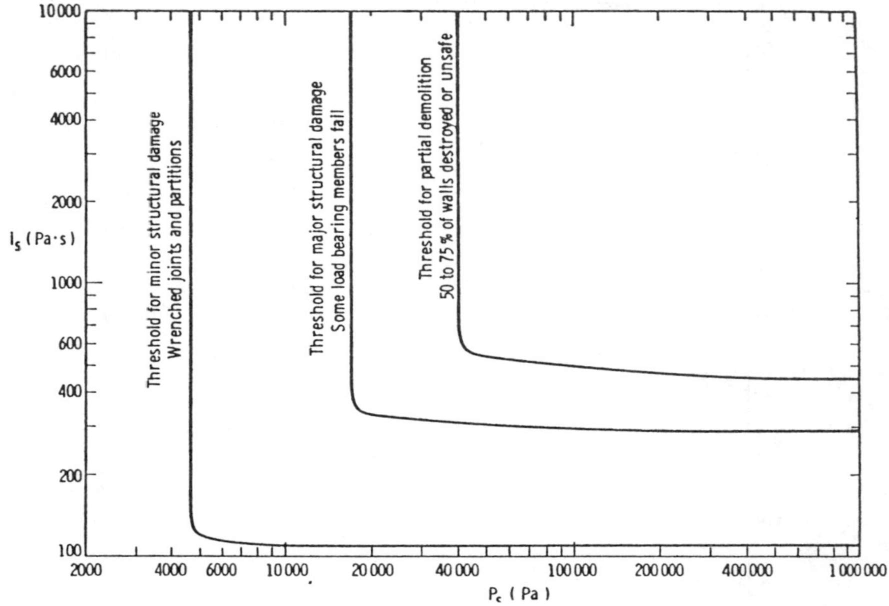

Figure 17. Impulse Versus Pressure Diagram for Constant Levels of Building Damage. (Reprinted with permission from ref. 15. Copyright 1983 Elsevier Science.)

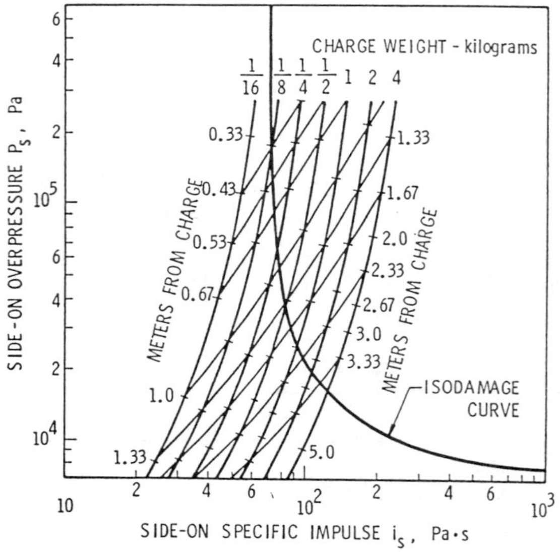

Figure 18. Illustration of Overlays to a P-I Diagram. Incident (Side-On) Overpressure and Impulse from Pentolite Spheres. (Reprinted with permission from ref. 15. Copyright 1983 Elsevier Science.)

effects of medium-sized HE-loaded projectiles against various types of conventional building walls appear in Fig. 19, from Ref. 23.

The P-i concept can also be used to collapse the results of a number of dynamic response calculations for structural elements into compact dimensionless design curves. A number of illustrations are given in Ref. 15, with one for blast-loaded beams with various boundary conditions appearing in Fig. 20. These curves give predictions of maximum dynamic bending strains and displacements for beams with a variety of boundary conditions. Details appear in Ref. 15, so we do not try to define all parameters here.

This method of presenting the topic of blast damage mechanisms was chosen primarily because it highlights the relationships between blast wave properties and structural response or damage. But, we hope that you now also know that the P-i or isodamage curves for structures can be useful design tools.

To Spall or Not to Spall. The amplitudes or peak reflected overpressures, of the reflected blast waves from high explosive detonations close to structures or structural elements can be very high. Figure 3 gives as a limit, for contact explosions of TNT, a pressure of P_r = 168,000 psi, while for an incident pressure P_s of 5000 psi, P_r = 61,000 psi. So, in addition to applying a very high and localized impulsive loading to the nearby structural surface, the explosion also applies compressive pressure pulses which peak very sharply to pressures well above compressive strengths of concretes, and even strengths of structural steels. Damage caused by the impacts, including damage from transmission and reflections of these intense waves, is termed "spalling" or "scabbing."

We should warn you that there is some confusion in definition of the two terms spalling and scabbing. In some civil engineering literature (see Ref. 24), spalling refers to scouring and ejecta damage to the loaded face of the structure or slab, while scabbing denotes wave-induced failures at the rear face of the loaded slab. But, this is not the usual physics definition, which instead uses the term spalling to cover all failures induced by intense wave transmission and reflections within solids. We use the more general physics definition. References 25-27 give good descriptions of the physics of shock transmission through solids, and spalling processes.

On the loaded side of a slab subjected to an intense reflected blast wave, a region of the slab will fail if the intensity of the compressive wave transmitted into the slab exceeds the dynamic compressive strength of the material. For an intense wave striking a thin concrete slab, the failure region can extend through the slab, and a sizeable area turned to rubble which can fall or be ejected from the slab. For a thicker slab or localized loaded area, spherical divergence of the stress wave can cause it to decay in amplitude within the slab so that only part of the loaded face side is crushed by direct compression.

The more common type of spalling failure of concrete occurs when (and where) the transmitted compressive wave reflects from the free surface back face of the slab as a tensile wave, and the head of the reflected tensile wave and tail of the transmitted compressive wave combine to produce net tensile stress exceeding the dynamic tensile strength of the concrete. This process is shown schematically in Figure 21 for the simplified case of a plane, triangular compressive

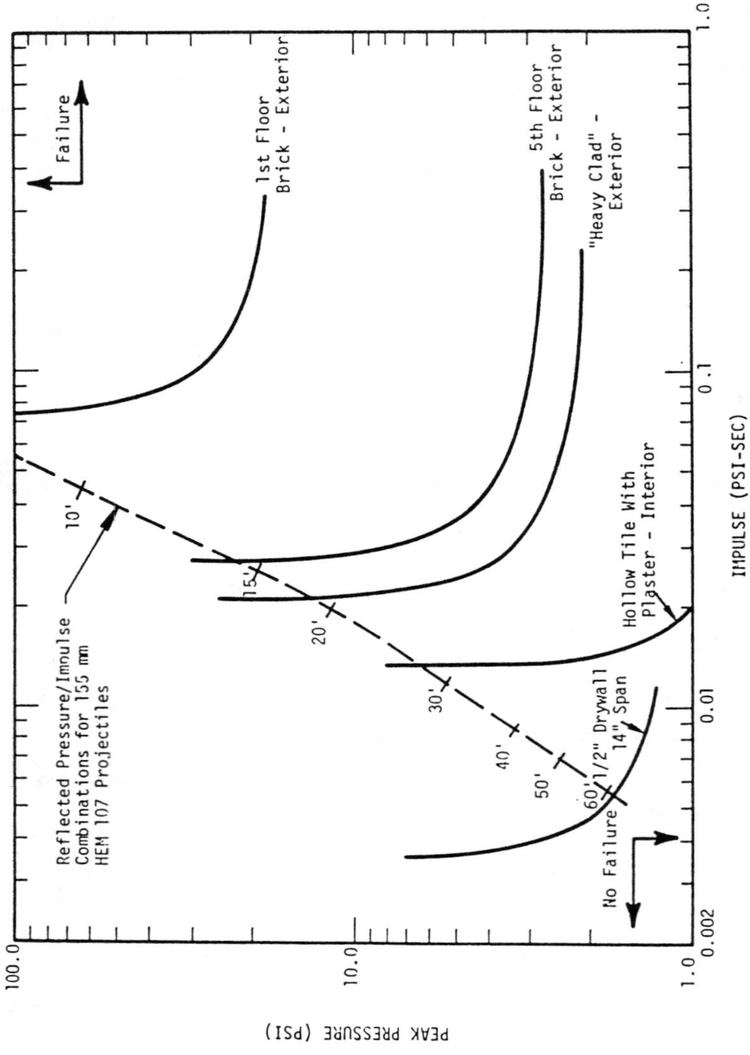

Figure 19. Isodamage Curves for Walls.(Ref. 23)

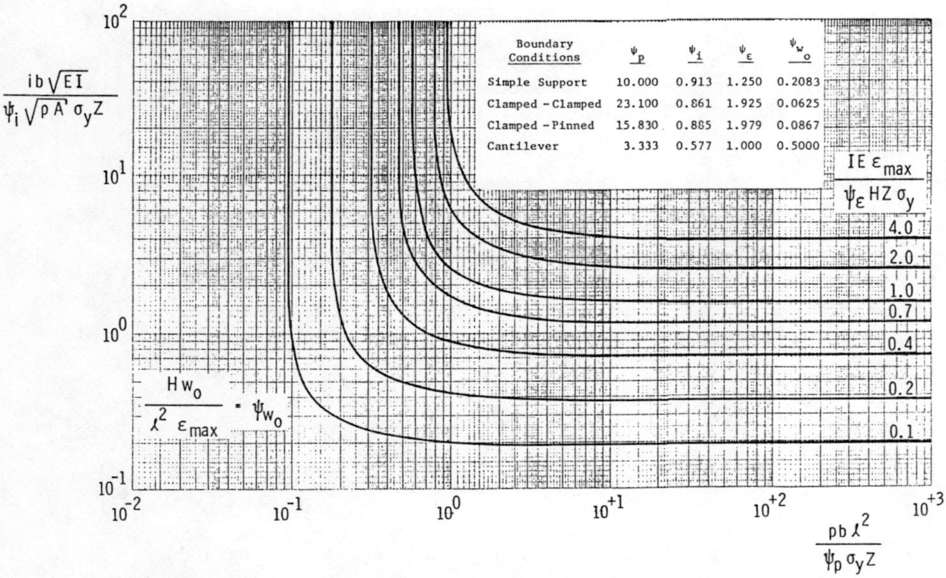

Figure 20. Elastic-Plastic Solution for Bending of Blast Loaded Beams. (Reprinted with permission from ref. 15. Copyright 1983 Elsevier Science.)

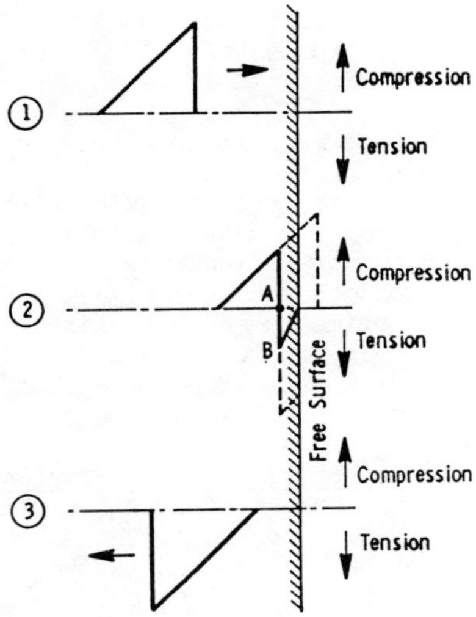

Figure 21. Stress Wave Reflection at a Free Surface in a Solid.

stress pulse reflected normally from a plane surface in a solid. The normal stress must be zero at the free surface, so a tension wave of a similar profile but opposite sign must start propagating in from the rear surfaces when the compressive front reaches this surface. The actual stress state shortly thereafter is shown in state 2 in Figure 21. When the tensile stress exceeds the tensile strength of the material, spall occurs on a plane parallel to the free surface. The normal stress then drops to zero again, and the process continues. In brittle materials weak in tension (such as concrete), it is possible for multiple spalls to occur before the reflected tensile waves decay enough to fall below the tensile strength.

For this simplified model of spalling, graphical boundaries have been determined for incipient spall for normally reflected air blast loading in Ref. 28, as shown in Figure 22. In this figure, terms not already introduced are defined as follows:

$$V = \sqrt{E/p} \qquad (26)$$

is the elastic dilatational wave speed in the solid,

H is wall thickness, and
σ_u is ultimate tensile strength of the wall material.

In preparing this figure, the authors of Ref. 28 assumed no wave attenuation through the wall thickness H, so P_r and i_r are the normally reflected blast loading parameters on the loaded side of the wall or slab.

Spalling can occur for quite strong materials such as structural steels and instances are shown in Refs. 25-27 for contact or near contact detonations. But of course it is more prevalent for weaker materials.

For complex composites such as reinforced concrete, the use of simple wave reflection analyses to predict spalling is quite suspect. So, several investigators have simply studied these thresholds experimentally. One of the most complete such studies is reported in Ref. 29. The author defined various damage categories for explosions near reinforced concrete walls, as in Figure 23. Then, he conducted a number of experiments and established scaled curves for various damage levels, as in Figures 24 and 25. The latter two curves can be used for quick estimates for both spalling and breaching of typical reinforced concrete wall panels.

<u>Shock Response Versus Quasi-Static Response for Internal Blast</u>. We noted earlier that internal detonations of high explosives within structures caused both initial and reflected shock loadings, plus longer term gas pressure loads called quasi-static pressures. Figure 11 is a reproduction of a pressure trace showing both phases of the loading.

Damage from internal blast is of course a function of the complete time history of the pressure loading. But, the duration of the shock phase of the loading is usually much shorter than duration of vented gas pressure loading, while the amplitude of the shock phase is much greater than peak quasi-static pressure. Quite often, the fundamental periods of walls or roofs are much longer than the shock

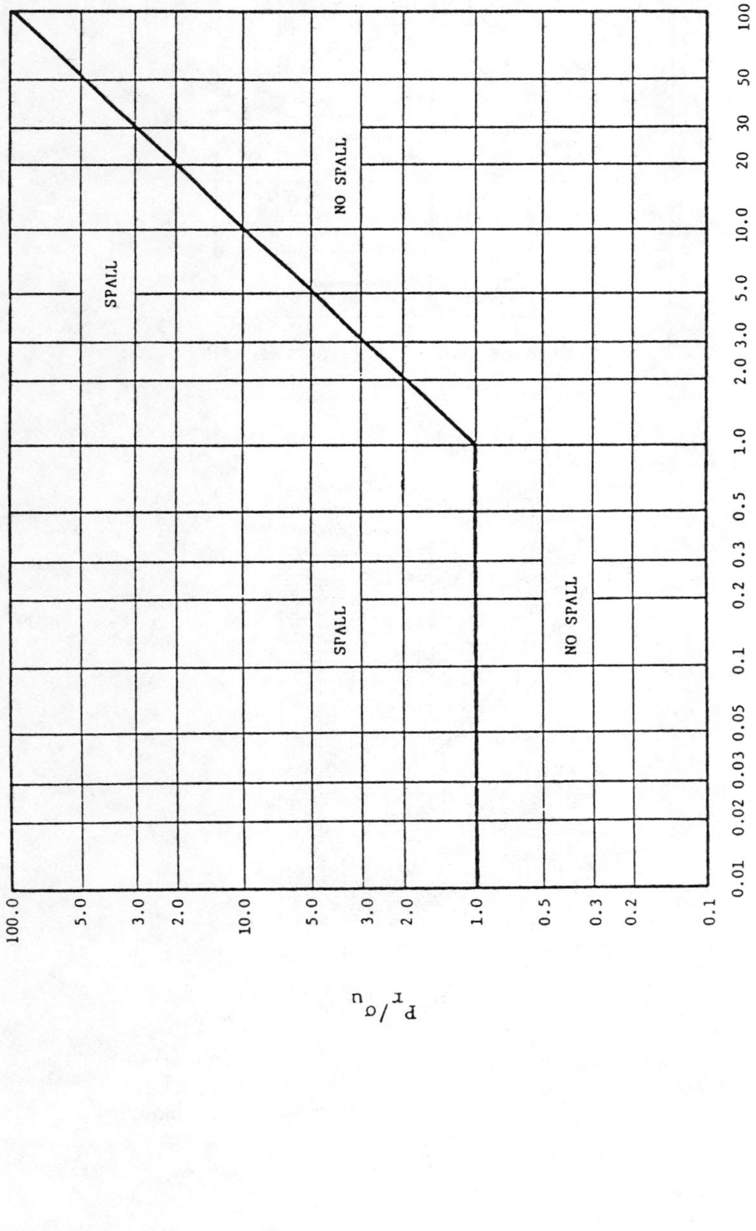

Figure 22. Spall Threshold for Blast Waves Loading Walls. (Ref. 28)

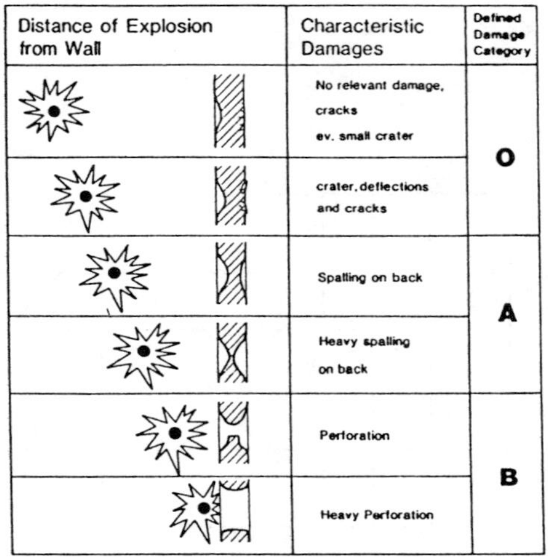

Figure 23. Definition of Damage Categories. (Ref. 29)

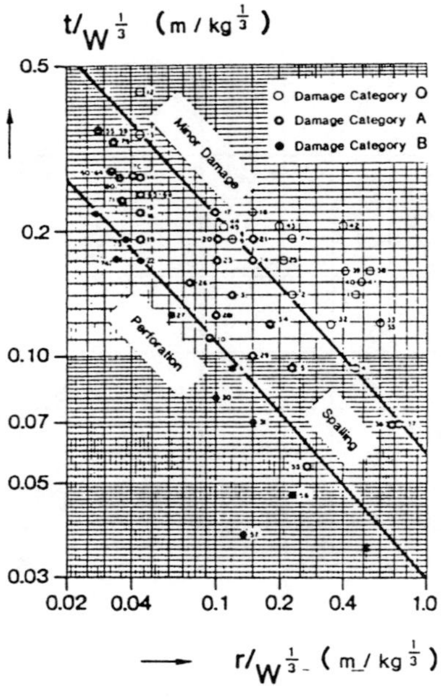

Figure 24. Damage to Reinforced Concrete Walls caused by Detonation of Uncased Explosives Charges. (Ref. 29) ($r/W^{1/3}$ = Scaled Distance; $t/W^{1/3}$ = Scaled Wall Thickness)

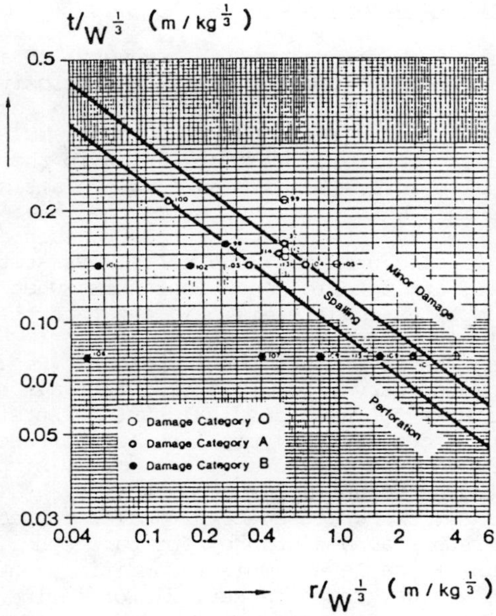

Figure 25. Damage to Reinforced Concrete Walls caused by Detonation of Cased Explosives Charges. (Ref. 29) ($r/W^{1/3}$ = Scaled Distance; $t/W^{1/3}$ = Scaled Wall Thickness)

loading phase, but much shorter than the gas loading phase. So, the structure then responds primarily to shock impulse, and to peak quasi-static pressure. With good venting, both phases may be significantly shorter than structural periods, in which case the total impulse, shock plus gas impulse, governs. As always, dynamics of the response plus these relative times <u>must</u> be considered in evaluating the relative importance of shock loading versus quasi-static loading.

Prediction of Blast Overpressure Outputs

<u>Air Shock Parameters</u>. There are available several "Standard" sets of curves of scaled air blast parameters for high explosive detonations in air. Such curves are always presented in scaled format using either the Hopkinson-Cranz scaling (Refs. 6, 15 or 30) or Sachs scaling (Ref. 5) discussed earlier. When presented in the more common Hopkinson-Cranz scaled form, it has been common practice to use charge weight W or mass M in place of the charge total energy E, and also to key the curves to a "standard" explosive such as TNT (Refs. 6, 15 or 30). It is also presumed, but not always stated, that standard (sea level) atmospheric conditions exist when the explosions occur. Final assumptions usually employed are that the charges are bare and of spherical geometry.

Some sources such as the tri-service manual (Ref. 30) include sets of blast parameter curves for spherical free-air explosions and separate sets of curves for hemispherical surface burst explosions. This is superfluous except at very small scaled distances, because the free-air curves can be used for both situations by simply using a higher effective charge weight for surface bursts.

We include a set of Hopkinson-Cranz scaled curves for blast wave properties versus scaled distance, for bare spherical TNT detonated in free air under sea level ambient conditions, as Figure 26. This set of curves was developed in Ref. 31 for inclusion in the revision to Ref. 30.

When using Figure 26 to predict blast wave properties for conditions other than bare, spherical TNT detonated away from a reflecting surface and at sea level ambient conditions, suggested adjustments are as follows:

1) Account for a surface or near-surface burst by first calculating a new effective free-air charge weight, W_e, as

$$W_e = (1.7 \text{ to } 2.0) \times W \tag{27}$$

The lower value is used for explosions on sand or soil, while the upper value is used for explosions which cause no cratering.

2) Account for high altitude ambient conditions by using correction factors based on Sachs scaling (Ref. 28). Below 5000 ft. altitude, these corrections are negligible.

3) Account for change in type of explosion by using a TNT equivalency factor, unless good test data are available

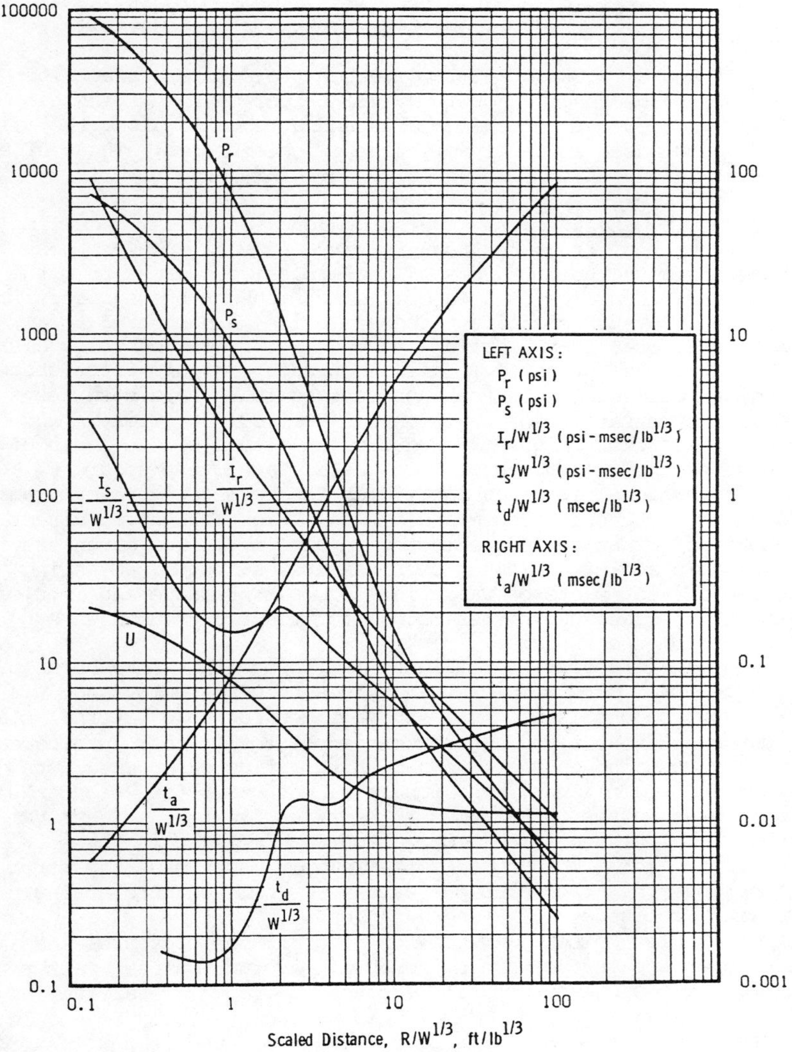

Figure 26. Airblast Parameters vs. Scaled Distance for a TNT Spherical Air Burst. (Ref. 31)

for the explosive. A rough estimate of TNT equivalency can be made based on relative heats of detonation of your explosive and TNT. This procedure is far from exact (see Ref. 6 and 32), but will allow you to at least estimate blast wave properties for other explosives.

4) See Ref. 28 for limited methods to predict effects of charge shape. For long cylindrical charges, blast is enhanced off the charge axis and attenuated along the axis, compared to equal weight spherical charges. These effects can persist to about 15 charge diameters.

Wall or Ceiling Shock Loads. The shock loads on walls or ceiling for explosions within structures usually vary appreciably over these surfaces, because the distances of the explosive sources from the surfaces are often less than lateral dimensions of the surfaces. So, the part of the surface nearest the explosive is subjected to a normally reflected shock, while all other parts feel an oblique shock sweeping over the surface. To help in predicting this first shock loading, experimental data for such surface loads have been curve-fitted, in preparation for revisions to Ref. 30. Figures 27 and 28 present these fits. Figure 27 requires knowledge of the angle of incidence of the oblique shock, and the side-on overpressure P_s. It then gives a multiplier which yields the reflected pressure on the surface at this incidence angle, $P_{r\alpha}$. Figure 28 gives directly the Hopkinson-Cranz scaled reflected impulse $i_{r\alpha}$, also given the incidence angle and peak side-on overpressure as inputs. By using these two curves, plots of variations of peak pressure and impulse over a wall surface can be estimated, for the first shock wave reflected from the surface.

Again referring to Figure 11, we see that the shock loads are, in general, more complex than this single pulse loading, with several reflected pulses. But, study of considerable internal blast data has shown that a good approximation to total shock loading can be made by assuming only second and third reflected shocks, with halving of the amplitudes (and impulses) each time (see Figure 29). Times between pulses are assumed to be twice the times of arrival for shocks calculated for explosive sources centered in the structure. If the total loading time $t_a + 4T_r$ is much less than structural period, then the three pulses can be combined into a single pulse with amplitude 1.75 P and duration T_r.

It is also common practice to integrate the pressures and impulses, over the surface areas, to obtain average values, rather than try and compute structural response to spatially-varying, as well as time-varying loads. But, this averaging procedure should be used cautiously for long walls or ceilings, because it can lead to serious underprediction of shock loads for part of the surface.

Quasi-Static Parameters. We noted earlier that the longer-term gas pressures which develop for explosions in vented or unvented structures can be characterized by three parameters; the peak quasi-static pressure P_{QS}, the duration t_{max} and the gas impulse i_g. For uncovered vents, reams of vented gas pressure data have been collapsed into scaled prediction curves and equations for these parameters in Ref. 18. We simply present that material here, as Figures

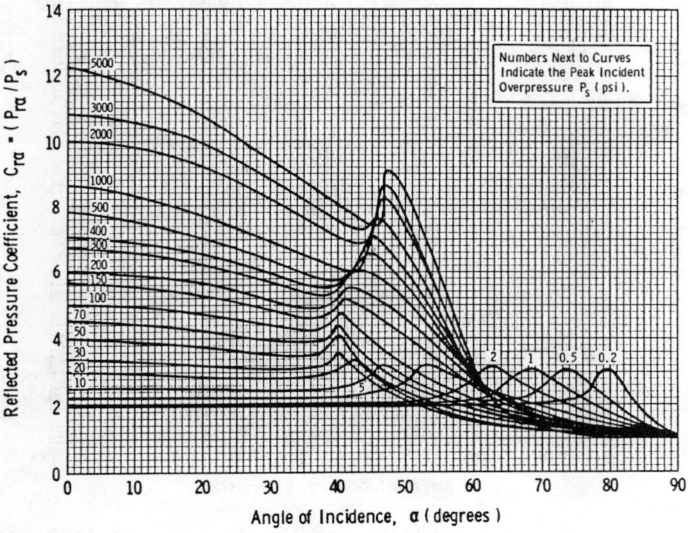

Figure 27. Reflected Pressure Coefficient Versus Angle of Incidence.

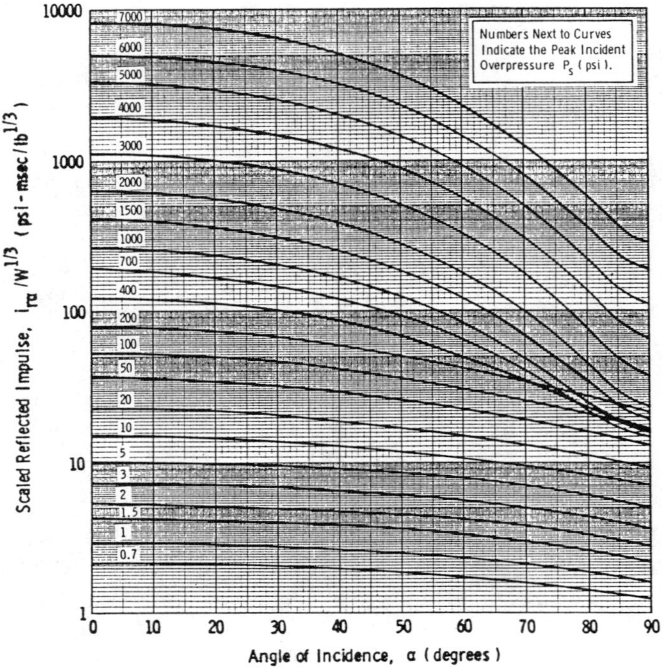

Figure 28. Reflected Scaled Impulse $i_{r\alpha}$ Versus Angle of Incidence α.

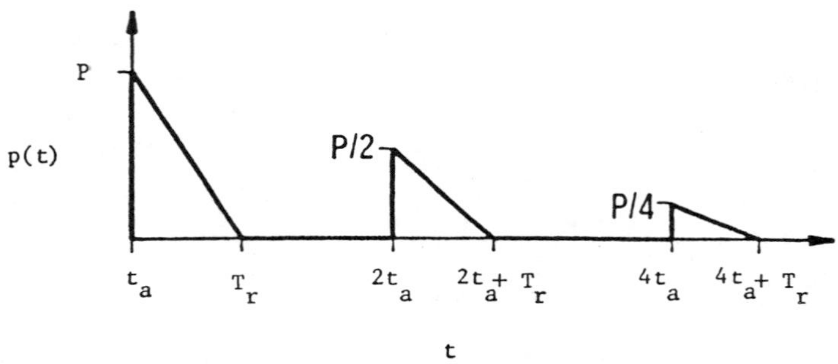

Figure 29. Schematic of Repeated Blast Loading.

30-32 and Tables I-III. Note that limits of applicability and standard deviations appear in the tables.

In most explosives safety structures, completely uncovered vent areas are unacceptable, either for environmental or security reasons. So, the vent areas have covers which may be quite light and frangible but have some inertia. Both analyses and testing have shown that even very light vent covers can significantly increase the duration and gas impulse for the gas pressure phase of internal blast loading. This work has been reduced to prediction curves which will appear in the revision to Ref. 30, as reported in Ref. 33. There are too many curves to reproduce here, but one is shown as Figure 33 to indicate its nature. The quantity γ is the specific weight of the vent panel, in lb/ft^2, the charge weight is in lb TNT equivalent, and room volume V is in ft^3, for this figure.

Table I. Summary of Peak $\bar{p}$ vs (W/p_oV) (Ref. 18)

$$\bar{p} = \frac{p_{QS} + p_o}{p_o}$$

$W/p_oV \leq 100$ $\qquad$ $\bar{p} = 1.336\ (W/p_oV)^{0.6717}$

Correlation Coefficient, r: 0.977

One Standard Deviation, σ_o : 1.164

$W/p_oV \leq 350$ $\qquad$ $\bar{p} = 1.336\ (W/p_oV)^{0.6717}$

Correlation Coefficient, r: 0.977

One Standard Deviation, σ_o: 1.262

$W/p_o > 700$ $\qquad$ $\bar{p} = 0.1388\ (W/p_oV)$

Correlation Coefficient, r: 0.896

One Standard Deviation, σ_o: 1.300

Containment and Venting Techniques

Containment Structure Concepts. In some types of safety facilities, it is either necessary or desirable to completely contain the effects of internal explosions. This requirement can arise because personnel, critical equipment, or critical operations must be located very near the facility, so one wishes to entirely eliminate blast emitted from the safety structure. A more stringent requirement requiring complete containment occurs in facilities for demilitarization of chemical munitions. Here, the extremely toxic

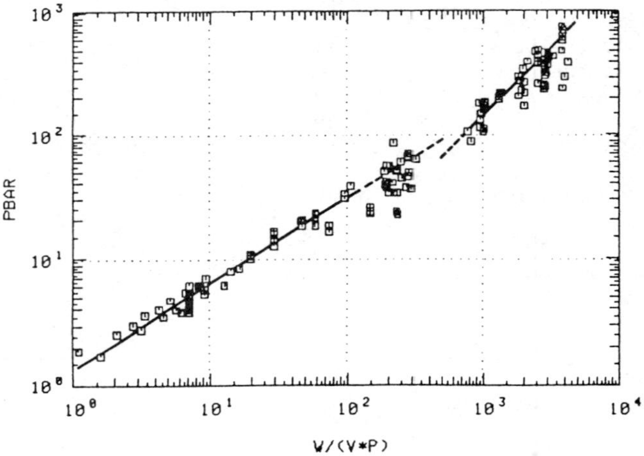

Figure 30. Reduced Pressure Versus Reduced Energy Density. (Reprinted with permission from ref. 18. Copyright 1983 Pergamon Press.)

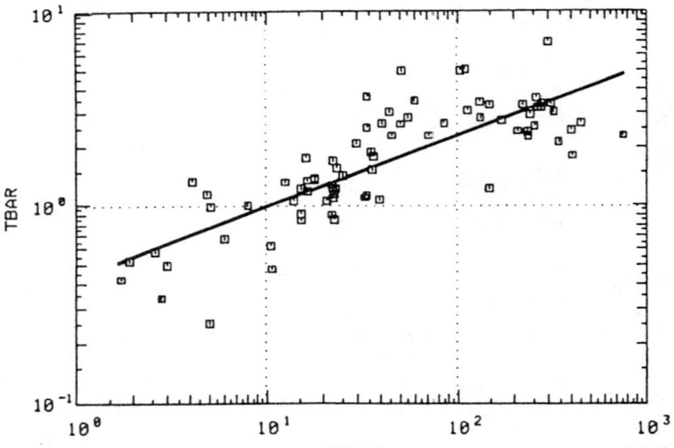

Figure 31. Reduced Duration Versus Reduced Pressure. (Reprinted with permission from ref. 18. Copyright 1983 Pergamon Press.)

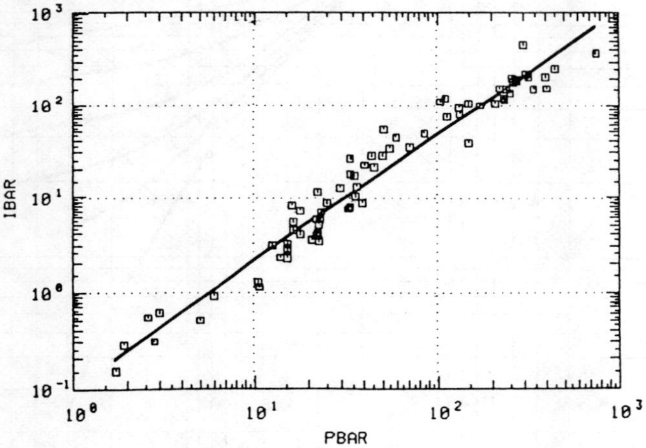

Figure 32. Reduced Specific Impulse Versus Reduced Pressure. (Reprinted with permission from ref. 18. Copyright 1983 Pergamon Press.)

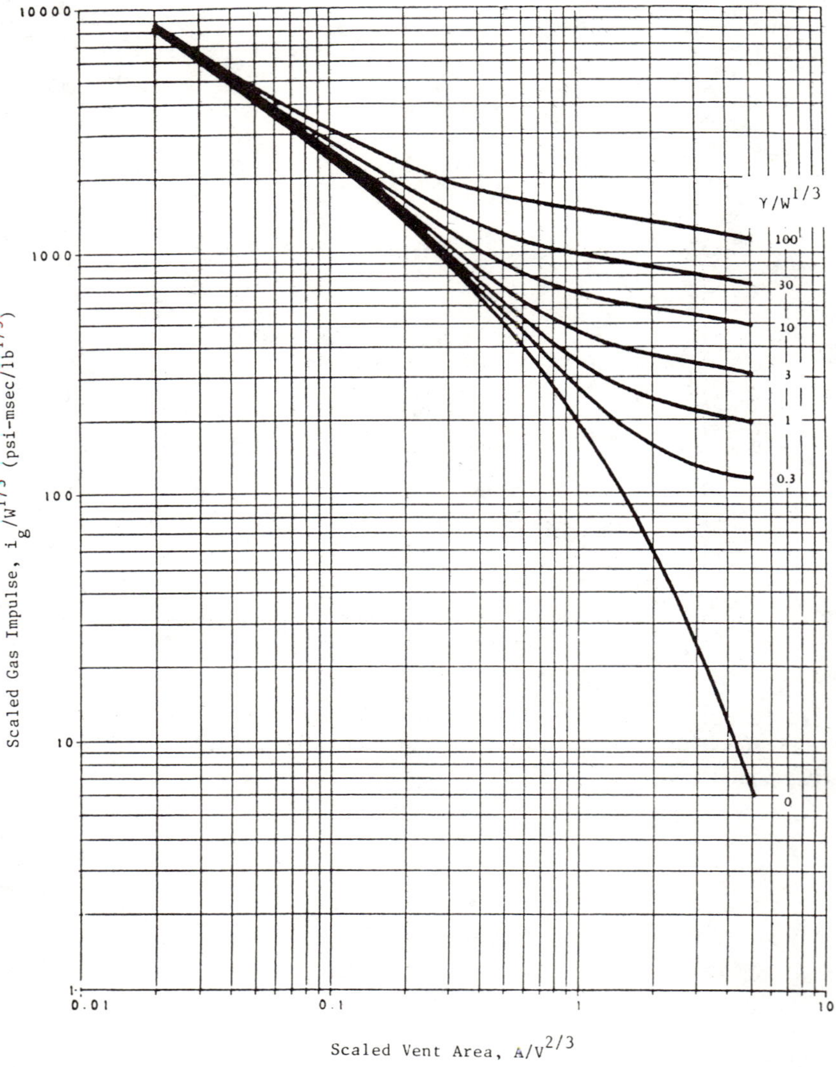

Figure 33. Gas Impulse Inside Structure with Frangible Panel. (W/V = 0.015, $i_e/W^{1/3}$ = 20) (Ref. 33)

Table II. Summary of $\bar{\tau}$ vs $\bar{p}$ (Ref. 18)

$$\bar{\tau} = \left(\frac{t\ a_o}{V^{1/3}}\right)\left(\frac{\alpha_{eff}\ A}{V^{2/3}}\right)$$

$$\bar{p} = \frac{p_{QS} + p_o}{p_o}$$

$$\bar{\tau} = 0.4284\ (\bar{p})^{0.3638}$$

Correlation Coefficient, r: 0.799

One Standard Deviation: $\sigma_o = 1.50$

Table III. Summary of $\bar{i}_g$ vs $\bar{p}$ (Ref. 18)

$$\bar{i}_s = \frac{i_g\ a_o\ \alpha_{eff}\ A}{p_o V}$$

$$\bar{p} = \frac{p_{QS} + p_o}{p_o}$$

$$\bar{i}_s = 0.0953\ (\bar{p})^{1.351}$$

Correlation Coefficient, r: 0.977

One Standard Deviation: $\sigma_o = 1.53$

nature of the chemical agents dictates the containment in the event of accidental detonation of explosive bursters during demil operations.

The size, shape and materials of construction depend on the function of the facility, the net explosive weight (NEW) for the worst-case accidental explosion in the facility, and other factors. Both reinforced concrete and steel have been used as materials, and shapes range from box (room) shaped, through horizontal and vertical cylinders to spheres. Generally, the room-shaped structures are most economically designed and constructed of reinforced concrete, while cylindrical and spherical shapes are most efficiently designed when made of steel.

In this keynote chapter, we give no details of containment structure configurations and designs. But, we note that Ref. 34

includes a comparison study for different cell configurations for chemical munitions demil operations.

Venting Techniques. The majority of explosive safety structures designed to mitigate or control the effects of internal explosions are vented in some fashion. The structures then attenuate or mitigate blast effects in adjacent bays or rooms, but do not completely contain these effects. Proper venting can significantly reduce or even eliminate gas pressure durations and impulses, and thus reduce total internal blast loads on safety structures. But, you are warned that venting is essentially totally ineffective in reducing internal shock loads.

Directional Venting. Most vented explosion safety structures are designed with blowout wall panels, entire walls, entire roofs, or even the entire roof and one wall. Other walls and roofs in the structure are designed to withstand a worst-case explosion without catastrophic failure. The explosion-proof parts of the structure provide some close-in blast protection, and hopefully complete protection from fragments and thermal radiation. But blast in the venting directions is not always attenuated compared to free-field blast and can even be enhanced in certain directions.

The most complete study of these directional venting effects for no vent covers is reported in Ref. 35. The results of scaled external blast tests in cubicles with various vent area ratios, $A/V^{2/3}$, from 0.020 through 0.77 and a variety of "loading densities" W/V are reported and presented for different vented cubicle configurations, including those with venting of the entire roof and one wall. Highly directional effects persist for some distances from these cubicles for some configurations. We have already noted a more recent report (Ref. 33) giving predictions for quasi-static loading parameters within directionally-vented cubicles with covers having various masses per unit area.

Many explosion safety structures utilize partially-buried designs, to minimize costs by providing earth support for blast-resistant walls and to prevent bay-to-bay propagation. Some of these structures are designed to vent relatively slowly through earth-covered or ground-covered roofs. Two such designs have been proof-tested with good internal and external blast instrumentation (Ref. 36 and 37).

For internal blast tests of a replica of a box-shaped, earth backed bay in the Pantex Plant at Amarillo, Texas, as in Figure 34, some venting occurred through the entranceway (which was not designed for containment), but the venting roof opened slowly and almost completely attenuated external blast waves venting through the roof.

A "Gravel Gertie" structure consists primarily of an earth-backed cylindrical reinforced concrete bay, with a deep gravel bed roof supported on a network of steel cables, as in Figure 35. In an internal blast test of the refurbished prototype for this type of structure, there was no blast venting from the simulated staging bays opening into the main cylindrical bay, and the main bay vented so slowly by upward displacement of the gravel roof that there was no measurable external blast. The slowly-moving gravel bed also

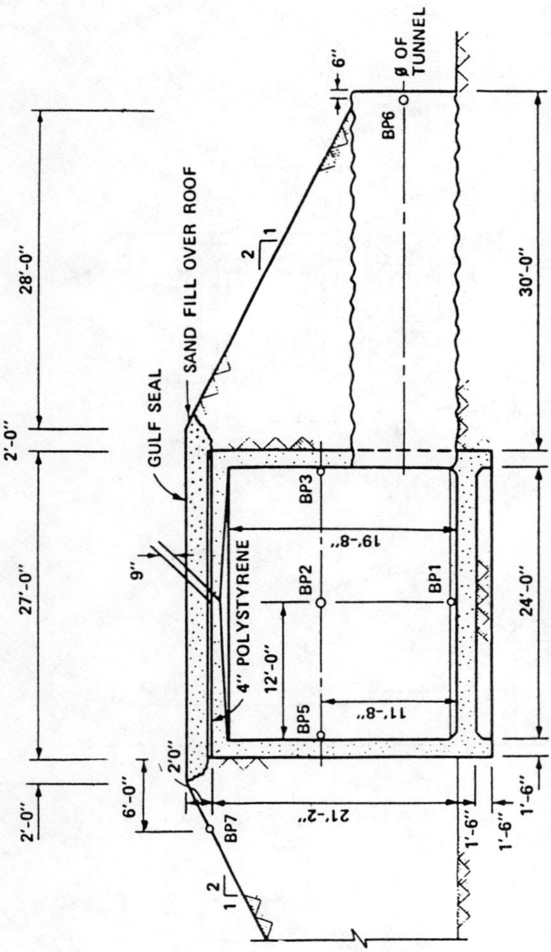

Figure 34. Elevation View of Replica of Partially Buried, Earth-Covered Bay.(Ref. 36)

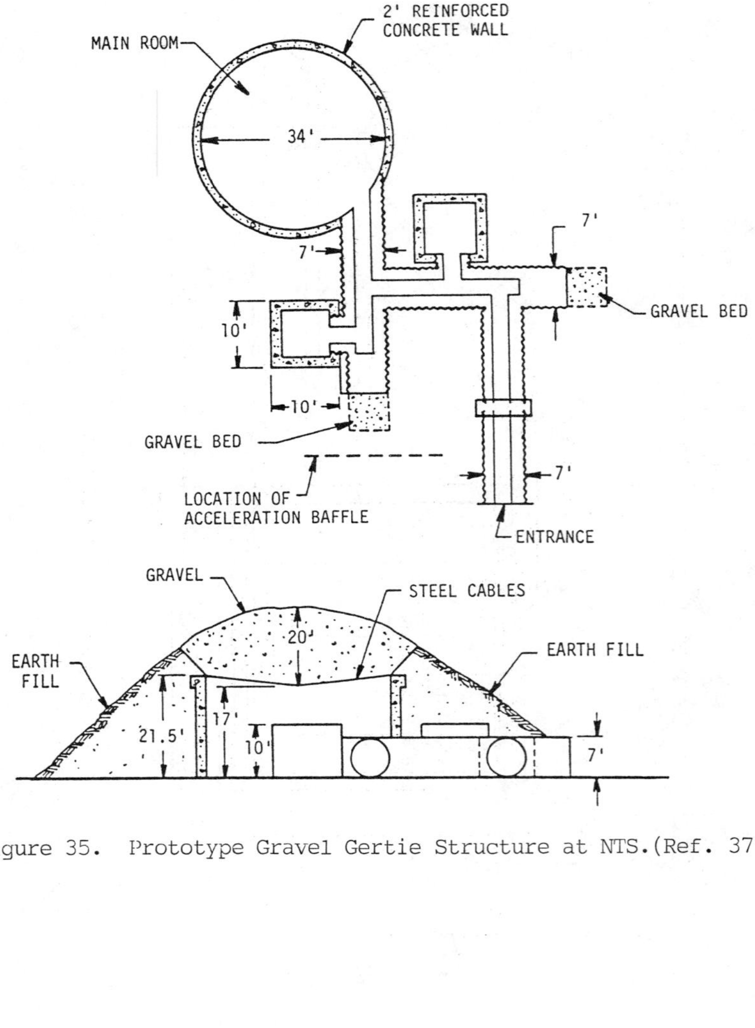

Figure 35. Prototype Gravel Gertie Structure at NTS.(Ref. 37)

proved to be an efficient dynamic filter for small toxic particles from the explosion.

Omnidirectional Blast Venting. During the period 1973-1977, Edgewood Arsenal sponsored an extensive program to evaluate the concept of steel explosion safety structures which were vented on all sides, or all sides plus roof. These structures, intended to be fabricated primarily using standard structural steel members, consisted of frameworks supporting multi-layered vent panels. They were termed "suppressive shields". The vent panels were all designed to attenuate air blast for explosions within the shields, and the layers in the panels were offset to prevent direct passage of fragments.

By the conclusion of the program, a number of designs had been built and tested, and proved quite effective. Methods for prediction of blast attenuation and fragment arresting capability of the designs were developed and verified.

There are numerous technical reports giving results of the extensive suppressive shields program, but they are well summarized together with design and analysis methods in a single design manual, Ref. 38. Seven shield designs have obtained safety approval from the Department of Defense Explosives Safety Board, and their specifications and construction drawings appear in an appendix to Ref. 38.

Typical sections through vent panels evaluated in the suppressive shields program are shown in Fig. 36, together with definitions of vent area ratios which were found to correlate with attenuation of transmitted blast waves.

The vent area ratio for a single layer structure is the vent area divided by the total area of the wall. The vent area ratio for a multi-layer structure is

$$\frac{1}{\alpha_e} = \sum_{i=1}^{n} \frac{1}{\alpha_i} \tag{28}$$

where α_e is the multi-layer and α_i is the single layer vent area ratio for an n-layer structure.

The vent area ratio for a perforated plate is simply

$$\alpha_i = A_{vi}/A_{wi} \tag{29}$$

where A_{vi} and A_{wi} are the vent area and wall area of the ith layer, respectively. For cubicles with a portion or all of a wall or roof missing, the vent area is the area of the opening and the appropriate value for α_e is the ratio of the open area to the total interior area of the cubicle.

Procedures for calculating vent area ratios for various structural configurations which have been used for suppressive shields are presented in Fig. 36. The procedures shown in Fig. 36 are believed to be self-explanatory, except possibly for the interlocked I-beams. The vent areas number 2 and 3 for this case are to take account of the two equal spaces b associated with each I-beam.

The expression for peak overpressure in psi outside a suppressive shield is (Ref. 38)

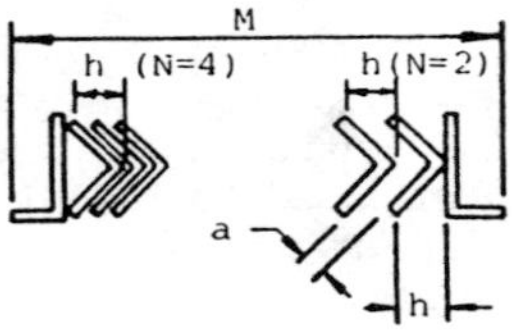

A_v = $n\ell a/N$
n = number of openings
ℓ = length of exposed element
h = projected width of angle
N = 2 for one opening per projected width h of angle; = 4 for two or more openings per width h
A_w = LM
L = length of wall
α_i = A_v/A_w
M = panel width

(a) Nested Angles

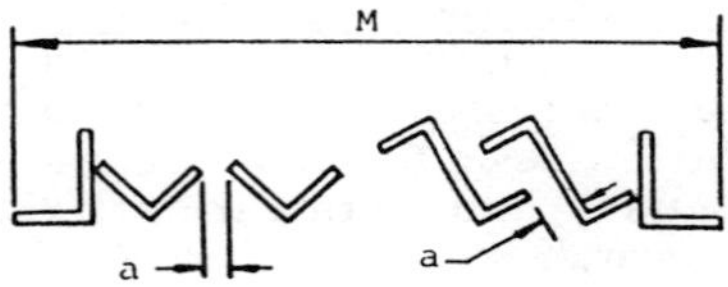

A_v = $n\ell a$
n = number of openings
ℓ = length of exposed element
A_w = LM
L = length of wall
α_i = A_v/A_w
M = panel width

(b) Side-by-Side Angles or Zees

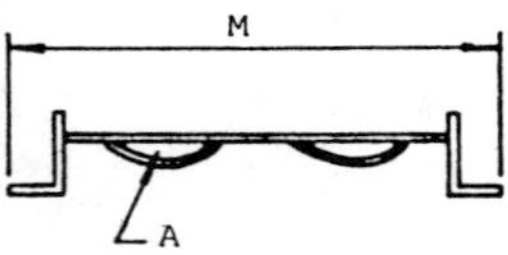

A_v = $nA/2$
n = number of louvres
A = open area of louvre
A_w = LM
L = Length of wall
α_i = A_v/A_w
M = panel width

(c) Louvres

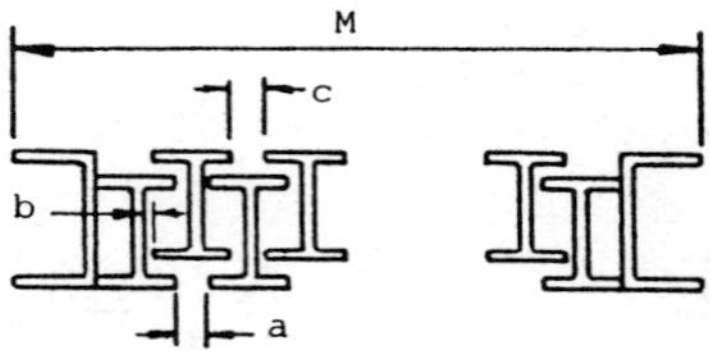

M = panel width
A_{v_1} = $2\ell n_a a$
A_{v_2} = A_{v_3} = $2\ell n_b b$
A_{v_4} = $2\ell n_c c$
n_a, n_b, n_c = number of openings a,b,c
ℓ = length of element
A_w = LM
L = length of wall
α_1 = A_{v_1}/A_w, α_2 = A_{v_2}/A_w, ...

(d) Interlocked I-Beams

Figure 36. Definition of Vent Area Ratios for Various Suppressive Shield Structural Configurations. (Ref. 38)

$$P_s = 957 \left(\frac{1}{Z}\right)^{1.66} \left(\frac{R}{X}\right)^{0.27} \left(\alpha_e\right)^{0.64} \qquad (30)$$

where

Z = Hopkinson-Cranz scaled distance, ft/lb$^{1/3}$

R = distance from center of explosive charge to point of interest, ft

X = characteristic length of structure, ft; side dimension for square structure; square root of plan area for rectangular structure; cube root of the volume for cylindrical structure

α_e = effective vent area ratio,

The limits for applicability of this equation are

$$2.93 \le Z \le 21.3$$

$$0.69 \le R/X \le 4.55$$

$$0.01 \le \alpha_e \le 0.13$$

and the expected error (standard deviation) is $\pm$ 19.9 percent.

The incident positive phase impulse in psi-ms outside a suppressive shield is given by (Ref. 38)

$$i_s = \left[218 \left(\frac{1}{Z}\right)^{0.98} \left(\frac{R}{X}\right)^{0.008} \left(\alpha_e\right)^{0.45}\right] W^{1/3} \qquad (31)$$

where W is in pounds of TNT and the other terms are as previously defined. The limits of applicability of this equation are

$$2.93 \le Z \le 15.0$$

$$1.16 \le R/X \le 4.55$$

$$0.008 \le \alpha_e \le 0.13$$

and the expected error (standard deviation) is $\pm$ 19.2 percent. Additional equations are available for specific panel designs with smaller standard deviations (Ref. 38).

Equations 30 and 31 apply to any vented panel configuration which has been tested (e.g., all safety approved shields) and to uniformly vented structures, i.e., structures vented in the same manner through all sides and the roof.

You are cautioned <u>not</u> to extrapolate equations (30) or (31) beyond their stated limits of applicability.

Risk Assessment Systems. Most of the techniques, design methods and applications in this book are deterministic. That is, some worst-case accident is assumed to happen, its effects are calculated to the best of our ability, and systems or structures are then designed to contain, suppress or mitigate the explosion accident effect. Only within about the last ten years have probabilistic methods been accepted in evaluation of potential explosion accidents by the Department of Defense in the United States. Such methods have a much longer history of development in Europe, particularly in Switzerland, and consequently are in much wider use there.

The primary document outlining risk assessment methods in the U.S. Department of Defense is a Military Standard, Ref. 39. This document requires a well-documented system safety program, based on risk assessment methods to be included in all new Department of Defense systems and facilities. Hazards analyses of the systems are mandated by this publication.

In Ref. 39, hazard severity categories are defined, as in Table IV.

Table IV. Hazard Severity Categories Defined in MIL-STD-882A (Ref. 39)

Category I - Catastrophic*. May cause death or system loss.

Category II - Critical*. May cause severe injury, severe occupational illness, or major system damage.

Category III - Marginal*. May cause minor injury, minor occupational illness, or minor system damage.

Category IV - Negligible. Will not result in injury, occupational illness, or system damage.

*Often expanded with subcategory B for effects on personnel, and subcategory A for effects on systems.

Ref. 39 suggests an initial qualitative hazards analysis early in systems design, with only general levels of hazard probabilities identified, in addition to severity categories. An example of such a qualitative ranking from Ref. 39 appears in Table V.

After initial design, in which serious hazards identified by a preliminary hazards analysis are hopefully eliminated or mitigated, Ref. 39 suggest a quantitative risk analysis. Here, specific numerical probabilities must be assigned for each damage category. Ref. 40 gives suggested levels, for the U.S. Army Production Base Modernization Program, as in Table VI.

Although risk analysis of new facilities is required by Ref. 39, the method of conducting the analysis is left quite open. The reference suggests fault hazard analysis, fault tree analysis, or sneak circuit analysis. Ref. 41 is an example of a thorough hazards evaluation and risk analysis for a new facility at Radford Army

Ammunition Plant. In the analysis, the probability levels of Table VI were used as requirements, and recommendations for changes made to subsystems which did not meet these requirements.

In Ref. 42, we see a review of the risk assessment methods used in Switzerland, and an application to assessing risks in solid propellant production. In the Swiss methods, one first defines individual risk r, as

$$r = W \times t \times \lambda \qquad (32)$$

where,

W = probability of event

t = probability of presence

λ = probability of fatal injury

Then, one evaluates collective risk R, as

$$R = \sum_{\text{persons}} r \qquad (33)$$

This process is shown schematically in Fig. 37.

Table V. Example of Qualitative Hazard Probability Ranking (Ref. 39)

Descriptive Word	Level	Specific Individual Item	Fleet or Inventory
Frequent	A	Likely to occur frequently	Continuously experienced
Reasonably Probable	B	Will occur several times in life of an item	Will occur frequently
Occasional	C	Likely to occur sometime in life of an item	Will occur several times
Remote	D	So unlikely, it can be assumed that this hazard will not be experienced	Unlikely to occur but possible
Extremely Improbable	E	Probability of occurrence cannot be distinguished from zero	So unlikely, it can be assumed that this hazard will not be experienced
Impossible	F	Physically impossible to to occur	Physically impossible to occur

Table VI. Design Goals for Probability Values for U.S. Army Production Base Modernization Program (Ref. 40)

Accident Category	Accidents per Facility-hrs	Accidents Per Man-hrs
IA	10^{-6}	-
IB	-	10^{-7}
IIA	10^{-5}	-
IIB	-	10^{-6}*
IIIA	10^{-3}	-
IIIB	-	10^{-6}*
IV	1	1

*Note: The sum of the probabilities of category IIB or IIIB occurring shall be 10^{-6} per man-hour or lower.

The total risk assessment process used by the Swiss is shown in Fig. 38. In Switzerland, an acceptable individual risk has been established to be 3×10^{-4}/year. It is interesting to note that this value is not far from that footnoted in Table VI, which converts to 10^{-3}/year. But, methods of calculating probabilities in Refs. 41 and 42 are quite different.

In use of risk assessment methods, you will find that the methodology for calculating overall risk probabilities is quite well defined. But, assigning realistic values to individual probabilities can be quite difficult, and a matter of personal opinion of the analyst. So, the analyst must have intimate knowledge of the system being evaluated, as well as all effects being considered, before he can make an acceptable risk assessment.

Biodynamics of Blasts

Human beings are surprisingly resistant to injury from air blast waves, compared to many structures. But, these waves can cause blast injuries to ears, lungs and other body parts; injuries from impact of debris on humans; and injuries caused by humans being tumbled or translated by the net transverse pressures and later striking the ground or some hard object. These three categories of blast injury are termed primary through tertiary injury.

In the United States, most of the studies on blast injuries to all types of mammals, including humans, have been done by the staff of the Lovelace Foundation. Their work is summarized in Refs. 43-45, and criteria given for primary air blast lethality levels for humans,

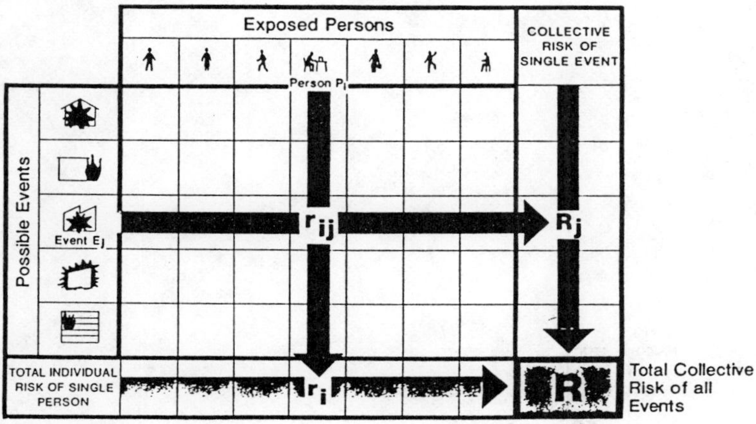

Figure 37. Risk Matrix for Swiss Risk Assessment Methods. (Ref. 42)

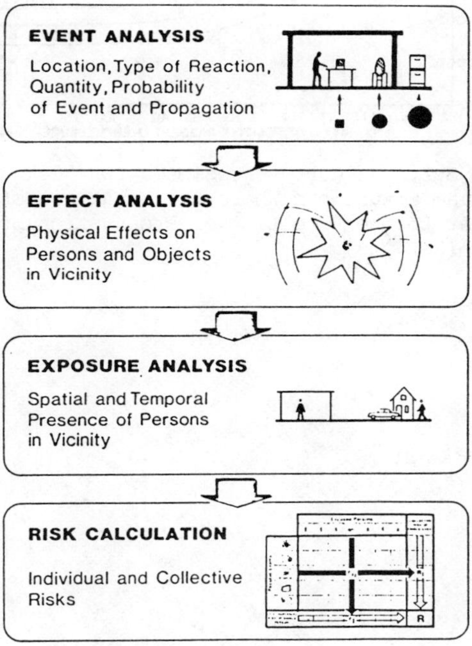

Figure 38. Steps of a Risk Analysis. (Ref. 42)

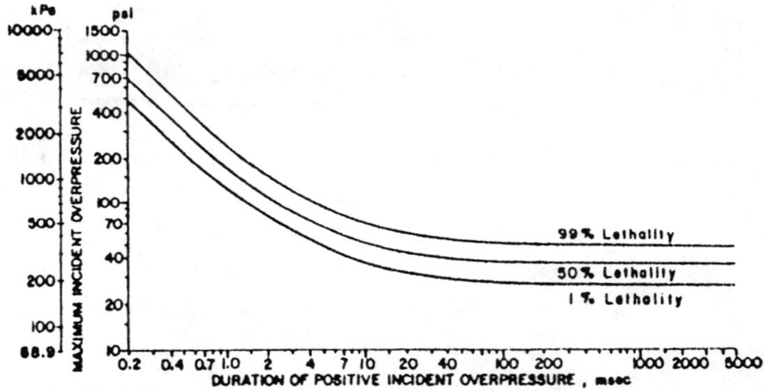

Figure 39. Survival Curves Predicted for 70-kg Man Applicable to Free-Stream Situations Where the Long Axis of the Body is Perpendicular to the Direction of Propagation of the Shocked Blast Wave. (Ref. 43)

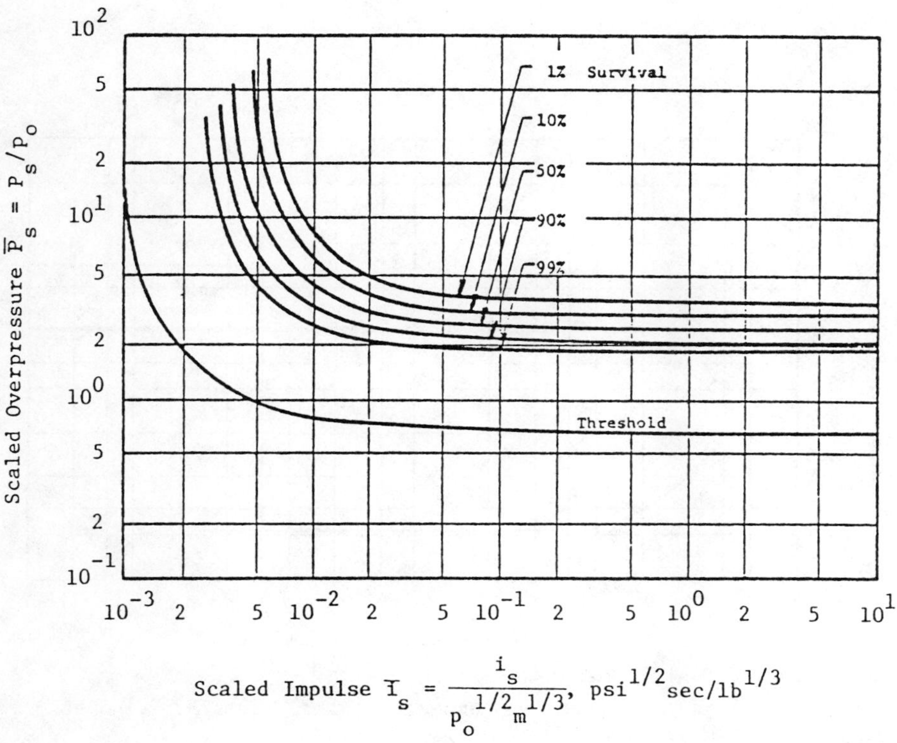

Figure 40. Survival Curves for Lung Damage to Man. (Ref. 46)

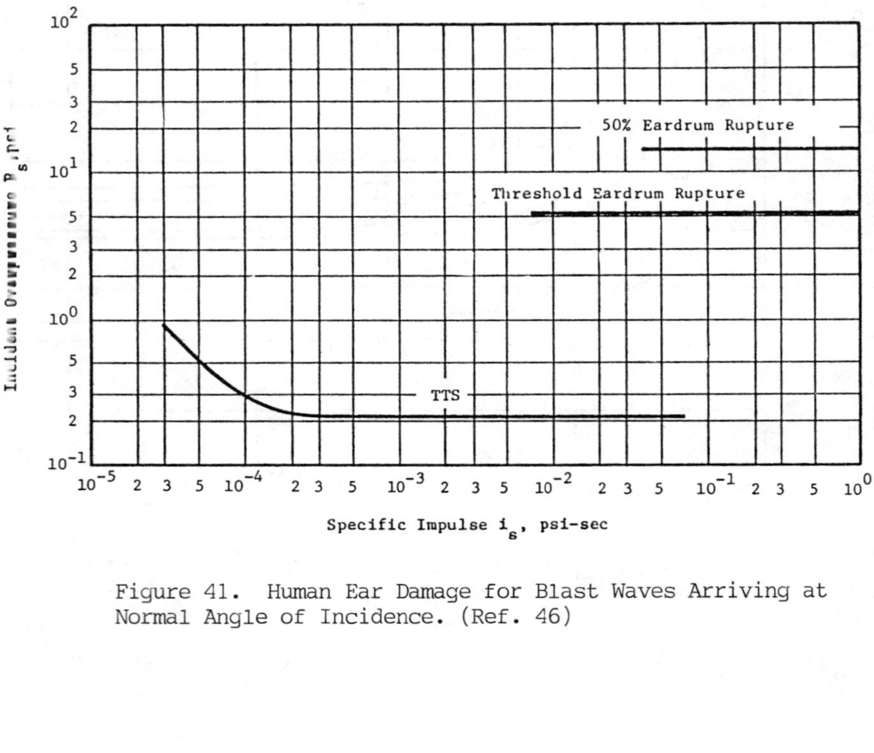

Figure 41. Human Ear Damage for Blast Waves Arriving at Normal Angle of Incidence. (Ref. 46)

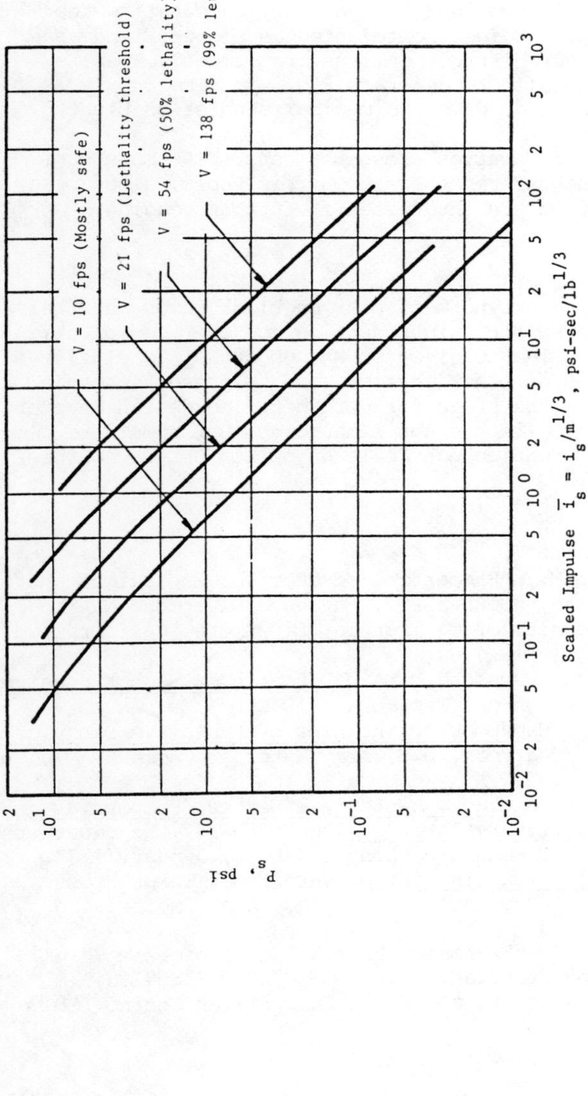

Figure 42. Lethality from Whole Body Translation. (Ref. 46)

as in Fig. 39, as combinations of incident blast wave overpressures and positive phase durations.

The Lovelace work was later converted in Ref. 46 to scaled curves for combinations of peak incident overpressure and positive phase specific impulse. These curves are reproduced here as Fig. 40. Hirsch's work (Ref. 44) can also be given as pressure-impulse combinations for ear injury, and this also was done in Ref. 46. The curves appear here as Fig. 41.

We do not treat secondary (fragment impact) effects in this chapter, but do present a set of curves for estimating injury from the tertiary effect of whole-body translation caused by blast diffraction and drag loading on a standing human. The curves were first reported in Ref. 46, and were developed by calculating the velocities to which human bodies (represented as short cylinders) would be accelerated under diffraction plus drag loads. Results were then collated and scaled to generate Fig. 42. The injury levels correspond to those observed in medical studies of blunt trauma.

Essentially all of the curves presented here, plus more complete discussions of and reference lists on this topic, also appear in Refs. 15 and 28, if you are interested in further reading.

Closure

It is hoped that this keynote chapter on blast waves and their effects will serve as a suitable introduction and overview of this topic. The author has tried to give you enough detail to clarify some of the fundamentals of blast physics, and to present material which will hopefully set the stage for more detailed design chapters to follow. The reference list is not exhaustive, but should be extensive enough and current enough to lead you to further sources for more detailed study.

Literature Cited

1. Dobratz, Brigitta M., "Properties of Chemical Explosions and Explosive Simulants," UCRL-52997, Lawrence Livermore National Laboratory, University of California, Livermore, California, March 1981.
2. Johansson, C.H. and Persson, P. A., Detonics of High Explosives, Academic Press, London and New York , 1970.
3. Engineering Design Handbook, Principles of Explosive Behavior, AMCP 706-180, Headquarters, U.S. Army Materiel Command, Washington, D.C., 1972.
4. Engineering Design Handbook, Explosives Series, Properties of Explosives of Military Interest, AMCP 706-177, Headquarters, U.S. Army Materiel Command, Washington, D.C., January 1971.
5. Baker, W. E., Explosions in Air, University of Texas Press, Austin, TX, 1973. (Reprinted by Wilfred Baker Engineering, San Antonio, TX 1983)
6. Swisdak, M. M., Jr., "Explosion Effects and Properties: Part I - Explosion Effects in Air," NSWC/WOL/TR 75-116, Naval Surface Weapons Center, White Oak, Silver Spring, MD. October 1975.

7. Glasstone, Samuel and Dolan, Philip J., "The Effects of Nuclear Weapons," United States Department of Defense and U.S. Department of Energy, 1977.
8. Hopkinson, B., British Ordnance Board Minutes 13565, 1915.
9. Cranz, C., Lehrbuch der Ballistik, Springer-Verlag, Berlin, 1926.
10. Sachs, R. G., "The Dependence of Blast on Ambient Pressure and Temperature," BRL Report 466, Aberdeen Proving Ground, MD, 1944.
11. Doering W. and Burkhardt, G., "Contributions to the Theory of Detonation," Translation from the German as Tech. Report No. F-TS-1227-IA (GDAM A9-T-4G), Headquarters, Air Material Command, Wright-Patterson AFB, Ohio, May 1949.
12. Brode, H. L., "Quick Estimates of Peak Overpressure from Two Simultaneous Blast Waves," Defense Nuclear Agency Report No. DNA45-3T, Contract No. DNA001-78-C-0009, R&D Associates, RDA-TR-107006-008, Marina Del Rey, California, December 1977.
13. Kennedy, W. D., "Explosions and Explosives in Air," in Effects of Impact and Explosion, M. T. White (ed.), Summary Technical Report of Div. 2, NDRC, Vol. I, Washington, D.C., AD 221 586, 1946.
14. Harlow, F. H. and Amsden, A. A., "Fluid Dynamics -- An Introductory Text," LA 4100, Los Alamos Scientific Laboratory, University of California, Los Alamos, New Mexico, February 1970.
15. Baker, W. E., Cox, P. A., Westine, P. A., Kulesz, J. J., and Strehlow, R. A., Explosion Hazards and Evaluation, Elsevier Scientific Publishing Co., Amsterdam, 1983.
16. Baker, W. E., Hokanson, J. C., Esparza, E. D., and Sandoval, N. R., "Gas Pressure Loads from Explosions Within Vented and Unvented Structures," Proceedings. The Interaction of Non-Nuclear Munitions with Structures, U.S. Air Force Academy, Colorado Springs, CO, May 1983.
17. Kingery, C.N., Schumacher, R., and Ewing, W.O., Jr., "Internal Pressure from Explosions in Suppressive Structures," Memorandum Report ARBRL-MR-02848, U.S. Army Ballistic Research Laboratory, Aberdeen Proving Ground, MD., June 1978.
18. Baker, W. E., Anderson, C. E., Jr., Morris, B. L., Wauters, D. K., "Quasi-Static Pressure, Duration and Impulse for Explosions (e.g., HE) in Structures," Int. J. of Mech. Sciences, Vol. 25, No. 6, pp 455-464, 1983.
19. Esparza, E. D., Baker, W. E., Oldham, G. A., "Blast Pressures Inside and Outside Suppressive Structures," Report EM-CR-76042, Edgewood Arsenal, Aberdeen Proving Ground, MD., 1975.
20. Hokanson, J. D., Esparza, E. D., Sandoval, N. R., and Baker, W. E., "Effects of Combustibles on Internal Blast Loads on the Damaged Weapons Facility, Final Report for Phase IV," Final Report on P.O. F0913400 for Mason & Hanger - Silas Mason Co., Inc., Pantex Plant, Amarillo, TX, Oct. 1982.
21. Biggs, J. M., Introduction to Structural Dynamics, McGraw-Hill Book Co., New York, NY, 1964.
22. Baker, W. E., Westine, P. S. and Dodge, F. T., Similarity Methods in Engineering Dynamics, Hayden Book Co., New Rochelle, NJ, 1973.

23. Zabel, P. H., Morris, B. L., Baker, W. E., and Westine, P. S., "Evaluation of Indirect Fire Munitions for Use in Built-up Areas," Technical Memorandum 15-82, U.S. Army Human Engineering Laboratory, Aberdeen Proving Ground, MD., Dec 1982.
24. Kennedy, R. P., "A Review of Procedures for the Analysis and Design of Concrete Structures to Resist Missile Impact Effects," Nuclear Engineering and Design, Vol. 37, 1976, pp. 183-203.
25. Davids, N. (ed.), International Symposium on Stress Wave Propagation in Materials, Interscience Publishers, New York, NY, 1963.
26. Kolsky, H., Stress Waves in Solids, Dover Publications, Inc., New York, NY, 1963.
27. Rinehart, J.S., Stress Transients in Solids, Hyper Dynamics, Santa Fe, NM, 1975.
28. Baker, W. E., Westine, P. S., Kulesz, J. J., Wilbeck J. S., and Cox, P. A., "A Manual for the Prediction of Blast and Fragment Loadings on Structures," DOE/TIC-11268, U.S. Department of Energy, Amarillo Area Office, Amarillo, TX, Nov. 1980.
29. Hader, H. "Effects of Bare and Cased Explosive Charges on Reinforced Concrete Panels," Symposium Proceedings, The Interaction of Non-Nuclear Munitions with Structures, U.S. Air Force Academy, CO, May 1983, pp. 221-226.
30. "Structures to Resist the Effects of Accidental Explosives," Department of the Army Technical Manual TM 5-1300, Department of the Navy Publication NAVFAC P-397, Department of the Air Force Manual AFM 88-22, Department of the Army, Navy and Air Force, June 1969.
31. Kingery, C. N., and Bulmash, G., "Airblast Parameters from TNT Spherical Air Burst and Hemispherical Surface Burst," Ballistic Research Laboratory Technical Report ARBRL-TR-02555, Aberdeen Proving Ground, MD., 1984.
32. Esparza, E. D., "Blast Measurements and Equivalency for Spherical Charges at Small Scaled Distances," Int. J. Impact Engng. Vol. 4, No. 1, pp. 23-40, 1986.
33. Tancreto, J. E. and Helseth, E. S., "Effect of Frangible Panels on Internal Gas Pressures," Minutes of the Twenty-First Explosives Safety Seminar, Houston, TX, Aug. 1984, Vol. I, pp. 365-394.
34. Whitney, M. G., Friesenhahn, G. J., Baker, W. E., and Vargas, L. M., "A Manual to Predict Blast and Fragment Loadings from Accidental Explosions of Chemical Munitions Inside an Explosion Containment Structure," Vols. I & II, Final Report, Contract No. DACA87-81-0099, April 1983.
35. Keenan, W. A., and Tancreto, J. E., "Blast Environment from Fully and Partially Vented Explosions in Cubicles," Technical Report R828, Naval Civil Engineering Laboratory, Pt. Hueneme, CA, Nov. 1975.
36. Volz R. D., and Kiger, S A., "An Evaluation of the Separated Bay Concept for a Munition Assembly Complex; An Experimental Investigation of the Department of Energy Building 12-64 Complex," Technical Report SL-83-6, U.S. Army Waterways Experiment Station, Vicksburg, MS, Sept. 1983.

37. Esparza, E. D., and Baker, W. E., "Blast Measurements for the Full-Scale Gravel Gertie Test at Nevada Test Site," Final Report, P.O. No. F5305800, Mason & Hanger, Silas Mason Co., Inc., Amarillo, TX, Dec. 1982.
38. "Suppressive Shields Structural Design and Analysis Handbook," HNDM-1110-1-2, U.S. Army Corps of Engineers, Huntsville Division, Nov. 1977.
39. "Military Standard System Safety Program Requirements," MIL-STD-882A, June, 1977.
40. "Safety, System Safety Program for Modernization and Expansion Projects," DRC-PBM Memorandum No. 385-3, Office of the Project Manager for Munitions Production Base Modernization and Expansion, Dover, NJ, April 1978.
41. Ewing, T. W., Cabbage, W. A., Hoover, J. E., and Kristoff, F. T., "Hazard Evaluation and Risk Analysis of the NC Thermal Dehy Facility - Bldg. 3507," Report No. HA-79-R-2, Hercules Aerospace Division, Radford Army Ammunition Plant, Radford, VA, Feb. 1979.
42. Merz, H. A., "Methods of Quantitative Risk Assessment: The Case of the Propellant Supply System," Minutes of the Twenty-First Explosives Safety Seminar, Houston, TX, Aug. 1984, pp. 1485-1506.
43. Bowen, I. G., Fletcher, E. R., and Richmond, D. R., "Estimate of Man's Tolerance to the Direct Effects of Air Blast," DASA 2113 Lovelace Foundation, Albuquerque, NM, October 1968.
44. Hirsch, F. S., "Effects of Overpressure on the Ear - A Review," DASA 1858, Lovelace Foundation, Albuquerque, NM November 1966.
45. "Prevention of and Protection Against Accidental Explosions of Munitions, Fuels and Other Hazardous Mixtures," Annals of the New York Academy of Sciences, Vol. 152, Art. 1, pp. 89-162.
46. Baker, W. E., Kulesz, J. J., Ricker, R. E., Bessey, R. L., Westine, P. S., Parr, V. B., and Oldham, G. A., "Workbook for Predicting Pressure Wave and Fragment Effects of Exploding Propellant Tanks and Gas Storage Vessels," NASA CR-134906, Nov. 1975. (Reprinted Sep. 1977.)

RECEIVED March 6, 1987

Chapter 2

Fragmentation Effects: An Overview

Michael M. Swisdak, Jr., and Joseph G. Powell, Jr.

U.S. Naval Surface Weapons Center, 10901 New Hampshire Avenue, Silver Spring, MD 20903-5000

The phenomena and the effects of fragmentation produced by the detonation of energetic materials are discussed. These include the formation of primary and secondary fragments, the determination of fragment velocity, fragment number/mass distributions, and fragment impact probabilities. In addition, origins and applications of fragment hazard criteria are discussed. Finally, these criteria are applied in the generation of standards for safe standoff distances from fragmentation sources.

The detonation of any mass of energetic or reactive material can produce serious fragment hazards in addition to the blast (air shock) environment. Fragments which are ejected as a result of a detonation can be classed as either primary or secondary, depending on their origin. Primary fragments have, as their source, material which is in intimate contact with the explosive. Such material might be the casing of an explosive-filled artillery shell, the body of a press used for compaction of powdered explosives, or the walls of a kettle used for melting explosives. These fragments are usually small in size and travel initially at velocities on the order of thousands of feet per second. Secondary fragments are structural components and objects, which while not in contact with the explosive, are sufficiently near to it that they could experience substantial accelerations. These fragments are somewhat larger in size than primary fragments and travel, initially, at velocities of hundreds of feet per second.

FRAGMENTATION PHENOMENON

The classical, naturally fragmenting munition consists of an explosive-filled cylindrical shell. The case is generally machined or cast from a steel alloy.

This chapter not subject to U.S. copyright
Published 1987 American Chemical Society

During the detonation process, the case begins to expand rapidly. When the case reaches about one to one and a half times its original diameter, it begins to fracture (1). The end result of the fracture process is the formation of fragments. The fragment sizes may range from very fine (dust-like, weighing grains or less) to very coarse (spear-like, weighing pounds) and may consist of a myriad of geometric shapes. The rate of expansion greatly depends on the type of explosive as well as the case material and the geometry of the munition.

FRAGMENT VELOCITY

During the fragmentation process, the velocity of the fragments goes through three distinct regions. Near the charge surface, the fragments are accelerated over a small distance, usually a fraction of a charge diameter, from zero to some maximum velocity. Once this maximum velocity is obtained, it is sustained over some distance by the explosion product gas pressure at and behind the shock front. This region of near constant velocity can extend out to as much as twenty charge diameters in some situations. Beyond this region, drag forces become predominant, and the velocity decreases exponentially with distance.

Generally, when a particular fragment's velocity is measured, either electronically or photographically, it is obtained over some known distance. This type of measurement yields an average velocity.

The initial velocity of the fragment can then be obtained from this measured average velocity through the use of the equations (2,3) given below:

$$V_i = V_{av}(e^x-1)/x$$

where

$$x=.05(\rho_{air})(A_{frag})(C_d)(R)/M$$

and

V_i = Calculated fragment initial velocity
V_{av} = Average fragment velocity over the distance R
ρ_{air} = Air density
A_{frag} = Average fragment presented area
C_d = Drag coefficient
R = Distance over which the average fragment velocity was measured
M = Fragment mass

(NOTE: ρ_{air}, A_{frag}, R, and M must be in consistent units such that x is dimensionless.)

If the fragment initial velocity is known, or has been calculated, then the fragment velocity at any distance R can be calculated with the following equation, assuming a straight line trajectory (gravity effects ignored):

$$V_R = V_i e^{-x}$$

where

V_R = Fragment velocity at distance R
V_i = Fragment initial velocity
x = As defined above

One of the more important parameters in the above equations is the drag coefficient, C_d. The drag coefficient for any fragment is a function of its shape and its Mach Number (velocity divided by sound speed). For regular fragments, like spheres or cubes, the drag coefficients are reasonably well defined. For irregular fragments, like those resulting from detonating bombs or disintegrating concrete walls, no two fragments have exactly the same shape. As a result, no two irregular fragments have exactly the same drag coefficient. Work has been done, however, with some degree of success, to characterize the drag coefficients for these irregularly shaped objects (4,5). Table I presents average drag coefficient data for these irregular fragments taken from these references. Unless otherwise noted, the drag coefficient is assumed to vary linearly between the entries shown. (Note: It must be realized that the drag data has been normalized to determine the drag coefficient. Some references, such as 4 and 5, have defined the drag coefficient in terms of $\rho_{air} * V^2$, whereas Table I makes the definition in terms of $0.5 * \rho_{air} * V^2$. The results in Table I are, thus, a factor of two larger than the values provided in References 4 and 5.)

TABLE I DRAG COEFFICIENTS FOR IRREGULAR FRAGMENTS

MACH NUMBER	DRAG COEFFICIENT
0	0.80
0.75	0.88
0.90	1.09
1.15	1.26
2.00	1.14
4.00	1.08
>4.00	1.08

Note: Drag coefficient varies linearly between Mach Number entries

In cases where large supersonic velocities are encountered ($M \geq 2$), the drag coefficient remains fairly constant and the data presented in the above-cited literature are adequate. However, in situations where fragment trajectories are being calculated, the fragment spends a great deal of its flight time (approximately 75%) in the subsonic regime where the variations in drag coefficient have a major effect.

A series of subsonic and supersonic wind tunnel tests were performed on regular and irregular fragments (6). Analysis of the data produced by these tests indicated that the drag coefficient for an unstable, randomly tumbling steel fragment correlates best with the ratio of maximum fragment presented area to average fragment presented area. When this correlation is made, the uncertainty in the drag coefficient is reduced by about 40%. This technique is described in more detail in a recent paper by McCleskey (7).

Predictions can be made for the initial velocity of fragments provided certain properties of the explosive material are known. The Gurney equation (8), or one of its many variations (9,10), is the most widely accepted method for predicting fragment initial velocity. The equations are slightly different for spheres and cylinders:

cylinder: $$V=(2E)^{1/2}[(C/M)/(1+0.5(C/M))]^{1/2}$$

sphere: $$V=(2E)^{1/2}[(C/M)/(1+0.6(C/M))]^{1/2}$$

where

V = Fragment initial velocity
$(2E)^{1/2}$ = Gurney constant (depends on explosive composition)
C = Explosive weight (per unit length of cylindrical case)
M = Case weight per unit length

The measured Gurney constants for the same material seem to vary from experimenter to experimenter. Those presented in Reference 8 differ from those presented in References 11 and 12. These differences are on the order of 5-10 percent. Table II is taken from data presented in Reference 8. If data from either Reference 11 or 12 had been used instead, it would have resulted in a difference in the initial velocity of about 5 percent.

TABLE II GURNEY CONSTANTS FOR VARIOUS EXPLOSIVES

EXPLOSIVE	GURNEY CONSTANT (ft/s)
TNT	7260
PENTOLITE (50/50 PETN/TNT)	8100
NITROMETHANE	7380
COMPOSITION B (60/40 RDX/TNT)	8210
RDX	8940
H-6 (44.8/29.5/21/4.7 RDX/TNT/Aℓ/wax)	8380

The above discussion pertains, principally, to primary fragments. Other techniques are currently available to estimate the velocity and range of secondary fragments.

One of these, based on a semi-empirical relationship taken from Reference 12, predicts secondary fragment initial velocity:

$$V=K\left[0.556(R_e/R) + 2.75(R_e/R)^2\right]$$

and

$$K=A_p R_e g_s/M$$

where

- V = Secondary fragment initial velocity, in/s
- R_e = Radius of spherical charge, inches
- R = Range from center of explosive charge to nearest face of secondary fragment, inches
- A_p = Area of secondary fragment presented to explosive, in^2
- g_s = Secondary fragment shape factor
  - = 2/3 for sphere
  - = $\pi/4$ for side-on cylinder
  - = 1 for end-on cylinder
- M = Mass of secondary fragment, $lb\text{-}s^2/in$

One problem for this equation is the rather narrow limits of applicability. These limits are:

$$1.5 \leq R/R_e \leq 6.0$$

and

$$0.18\ \frac{lb\text{-}s}{in^3} \leq \frac{V}{K} \leq 2.0\ \frac{lb\text{-}s}{in^3}$$

In addition to its rather narrow limits of validity, this expression is strictly applicable only to spherical charges of Composition B. However, until further work is completed, this equation represents the best method available for predicting secondary fragment velocities.

Huang (13) describes a methodology for predicting secondary fragment debris ranges. He has developed a computer program, MUDEMIMP (Multiple Debris Missile Impact Simulation) that determines debris hazards by calculating the accumulated number of hazardous debris missiles at various impact ranges. The program employs a probabilistic approach by utilizing Monte-Carlo sampling techniques to assess the effects of variations and uncertainties on the debris launch characteristics.

FRAGMENT SIZE

One standard method for predicting fragment size, is a formula which relates the fragment mass and the fragment shape factor, or ballistic density (14):

$$m=kA^{3/2}$$

where

m = Fragment mass
k = Ballistic density
A = Fragment presented area

Another method, suggested by Porzel (15), is similar:

$$m=BALD$$

where

m = Fragment mass
B = Fragment shape factor; for irregular fragments 1/3 is a good estimate
A = Fragment area
L = Fragment length in direction of motion
D = Fragment density

FRAGMENT NUMBER

Various approaches are available for calculating the number of fragments with a mass greater than a given mass. Many of these approaches are compared in Reference 16. Two of the more popular are those proposed by Mott (14,17,18) and Porzel (19):

Mott $N(>m)=N_o\exp(-m/\mu)^{\gamma}$

Porzel $N(>m)=N_o\exp(-L/L_1)$

where

$N(>m)$ = Number of fragments of mass greater than m
N_o = Constant = total number of fragments (Note: N_o is not the same for both distributions)
m = Fragment mass
μ = Average fragment mass
γ = 1, 1/2, or 1/3, depends on
  μ = Average mass for γ = 1
  2μ = Average mass for γ = 1/2
  6μ = Average mass for γ = 1/3
L = Fragment length
L_1 = Characteristic fragment length

Unfortunately, there is no concensus as to which value, $\gamma = 1$, 1/2, or 1/3, applies to various fragmentation processes. The usual recommendation is to plot the data and observe which value of γ best fits the data.

HIT PROBABILITY

When fragments (either primary or secondary) are ejected, it is often necessary to calculate the probability of their impacting a particular target. Work by Klein (20) and Hackett (21) gives the hit probability equation as:

$$P = 1 - \exp(-qA_T)$$

where

P = Probability of hit
q = Areal density of fragments (number of fragments per area) at the range and direction of target
A_T = Area of target

(NOTE: q and A_T must be in consistent units.)

FRAGMENT HAZARD CRITERIA

The Department of Defense Explosives Safety Board (DDESB) defines a hazardous fragment as one with an impact energy of 58 ft-lb (79 joules) or greater. The DDESB also defines a hazardous fragment areal density as one hazardous fragment per 600 ft^2 ($56m^2$) (22). The origins of these criteria are not well established. Freund (23), presents an interesting synopsis of the history. The following discussion is excerpted from his paper.

A recent DDESB technical summary relating to fragment and debris hazards gives the areal density of injurious fragments considered acceptable under the current U.S. standards as one such fragment per 600 ft^2 of surface area, corresponding to an injuring probability of about one percent (20). (Authors note: Using the value of 6.2 ft^2 for the area (A_T) of a man (target), and 1/600 ft^2 to be the areal density of fragments (q), the hit probability equation cited above gives the hit probability to be: $P = 1 - \exp(-6.2/600) = 0.01$ or 1%). The one percent "acceptable" injury probability figure cited appears to have been chosen arbitrarily as a convenient one; no objective rationale for its acceptance has been found other than its prior acceptance in the U.K. and NATO countries for the 10-year period prior to the time that it was adopted by the DDESB, at its 260th meeting on 14 April 1971.

The 58 ft-lb criterion appears to have been borrowed initially from German army doctrine at the beginning of the present century (24). In its crudest form, this criterion stated that missiles with less than 58 ft-lb of kinetic energy do not kill, and that those with more than 58 ft-lb do kill. During World War II, the criterion of a missile with weight and velocity sufficient to give it 58 ft-lb of kinetic injury was used in practice. Although it was generally recognized that the adoption of the 58 ft-lb value was arbitrary, it was much more practical than using the penetration of pine boards or other inanimate objects for the purpose (25). Selection of the 58 ft-lb criterion was substantiated by the work of Gurney (26). The criterion was also in general agreement with the work of McMillen and his associates (27). Reference 25 sums up the situation by stating that "...while his 58 ft-lb figure...has not been fully substantiated as a fair criterion, it is well supported and is definitely superior to pine boards. No doubt, under optimal conditions, a missile with considerable less energy than 58 ft-lb can produce a serious wound, but on the average it is probable that this amount of energy will insure a casualty."

FRAGMENT HAZARD RANGE STANDARDS

The DDESB sets/defines minimum fragment distances to protect personnel in the open. Quoting from their standard (22) "...The minimum distance for protection from hazardous fragments will be based on the debris producing characteristics of the Potential Explosion Site (PES) and the population density of the Exposed Site (ES). For populous locations, the minimum distance will be that distance at which fragments, including debris from structural elements of the facility or process equipment, will not exceed a hazardous fragment density of one hazardous fragment per 600 square feet ($56m^2$). If this distance is not known the following shall apply:

(1) For 100 lbs NEW (45 kg NEQ) or less of demolition explosives, thin-cased or low fragmentation ammunition items, bulk high explosives, pryotechnics, and in-process explosives of Class/Division 1.1, the minimum distance to exposure listed above will be 670 ft (204m)... .

(2) For all types of Class/Division 1.1 in quantities of 101 to 30,000 lbs NEW (46 to 13,600 kg NEQ), the minimum distance will be 1250 ft (380m), unless it can be shown that fragments and debris from structural elements of the facility or process equipment will not present a hazard beyond the distance specified. For items that have been evaluated adequately, a different minimum distance...may be used.

(3) For public traffic routes that are not possible sites for future targets and for other exposures permitted at public traffic route distances, ...fragment...distance minima for Class/Division 1.1 may be reduced to 400 ft... .

For sparsely populated locations, the minimum fragment distance can be reduced to 900 ft (270m) if certain specific conditions exist as follows:

(1) No more than 25 persons are located in any sector bounded by the sides of a 45 degree angle, with the vertex at the PES, and the 900 ft (270m) and 1,250 ft (380m) arcs from the PES, and

(2) The NEW of the PES does not exceed 11,400 pounds (5,170 kg)."

LITERATURE CITED

1. Design for Terminal Effects, Ordnance Corps Pamphlet, ORDP 20 245, May 1957.

2. Joint Munitions Effectiveness Manual, Test Procedures for High Explosive Munitions, NAVAIR 00 130 ASR21, 8 Dec 1970.
3. Static Fragmentation Tests of High-Energy Munitions, U.S. Army Test and Evaluation Command, Test Operations Procedures, 31 Jan 1980.
4. Dunn, D. J. and Porter, W. R., Air Drag Measurements of Fragments, BRL 915, Aug 1955.
5. Calvert, J. J., Ballistic Tables for Fragments, NAVORD Report 5134, Oct 1956.
6. Daniels, P., McDonald, J., and Morgan, V., Subsonic, Transonic, and Supersonic Drag Characteristics of Nine Shape Categories of Warhead Fragments, NSWC TR 81-112, May 1981.
7. McCleskey, F., Drag Coefficients for Irregular Fragments, Twenty-Second Explosives Safety Seminar, Anaheim, CA, Aug 1986.
8. Gurney, R. W., The Initial Velocities of Fragments from Bombs, Shells, and Grenades, BRL Report 405, Sep 1943.
9. Jacobs, S., The Gurney Formula: Variations on a Theme by Lagrange, NOL TR 74-86, Jun 1974.
10. Randers-Person, G., An Improved Equation for Calculating Fragment Projection Angles, Proceedings of the Second International Symposium on Ballistics, American Defense Preparedness Association, 1975.
11. Petes, J., Blast and Fragmentation Characteristics, Prevention of an Protection Against Accidental Explosions of Munitions, Fuels, and other Hazardous Materials, Annals of the New York Academy of Sciences, Vol 152, Art 1, pages 1-913.
12. Suppressive Shields Structural Analysis and Design, Handbook, Report No. HNDM-1110-1-2, U.S. Army Corps of Engineers, Huntsville Division, Huntsville, AL, Nov 1977.
13. Huang, L.C.P., Prediction of Debris Hazards from Explosions in Buildings, Twenty-First Explosives Safety Seminar, Houston, TX, Aug 1984.
14. Johnson, C. and Moseley, J., Preliminary Warhead and Terminal Ballistics Handbook, NWL TR 1821, Mar 1964.
15. Porzel, F. B., Technology Base of the Navy Explosives Safety Improvement Program, Nineteenth Explosives Safety Seminar, Los Angeles, CA, Sep 1980.
16. Hokanson, J., Vargas, L., Whitney, M., Moseley, P., and Cardinal, J., Fragment and Debris Hazards from Accidental Explosions, NSWC TR 85-114, 13 Jul 1981.
17. Mott, N. F., A Theoretical Formula for the Distribution of Weights of Fragments, AC 3642 (British), Mar 1943.
18. Sternberg, H. M., Fragment Weight Distributions from Naturally Fragmenting Cylinders Loaded with Various Explosives, NOL TR 73-83, 12 Oct 1973.
19. Porzel, F., Petes, J., Tondo, D., Smith, C., Freund, D., and Swisdak, M., Naval Explosives Safety Improvement Program (NESIP): Summary and Status through 30 June 1976, NSWC TR 81-27, Mar 1981.
20. Klein, P. F., Fragment and Debris Hazards, Technical Paper 12, Department of Defense Explosives Safety Board, Jul 1975.

21. Hackett, O. F. and Peterson, R. O., Missile Hazards from Explosions in Ships, Nineteenth Explosives Safety Seminar, Los Angeles, CA, Sep 1980.
22. Department of Defense Ammunition and Explosives Safety Standards, DOD 6055.9-STD, Jul 1984 (First Amendment, Change 1, 19 Aug 86
23. Freund, D., Origin and Subsequent Modification of Explosive Safety Quantity-Distance (ESQD) Standards Special Reference to Naval Weapons--Volume 1, DTNSRDC/SD-78-4, May 1978.
24. Rohne, H., Schiesslehre fur Infanterie, 1906.
25. Coates, J. B., Wound Ballistics, Office of the Surgeon General, Department of the Army, 1962.
26. Gurney, R. W., A New Casualty Criterion, BRL Report 498, Oct 1944.
27. McMillen, J. J. and Gregg, J. R., The Energy Mass and Velocity Which is Required of Small Missiles in Order to Produce a Casualty, Medical Casualties Report 12, National Research Council, Division of Medical Sciences, pp. 93-94, 1945.

RECEIVED April 24, 1987

Chapter 3

Architectural Standard Details for Army Ammunition Plants

Richard W. Sime

Black & Veatch, Engineers-Architects, P.O. Box 8405, Kansas City, MO 64114

The procedure for designing facilities for Army Ammunition Plants (AAP) should be based on increased safety, and reduced maintenance, energy consumption, and costs. The primary objective for development of the Architectural Standard Details is to enhance safety and achieve uniformity of design. The other benefits obtained are possible by-products. Black & Veatch was engaged to develop details for use in design and construction of buildings in which nitroglycerin, nitrocellulose, and single base and multibase propellants are manufactured. This paper discusses the objectives, background, construction design requirements, use of standard details, typical details, and the procedure for making future changes to conform to advances in technology, architectural practice, or changes required by actual field performance of certain standard details.

For many years the Government has constructed Army and Navy ammunition plants throughout the country in association with commercial producers. Manufacturing plant structures were designed incorporating specific requirements imposed by plant operating contractors for the particular function of a structure and specific requirements of the type of explosive or propellant end product. Architectural details were developed by plant operating contractors, engineering firms engaged in plant design, and supervising government agencies. Many of the architectural details were developed with safety considerations specifically in mind and were originated by plant designers in order to protect plant personnel from the effects of explosives manufacturing accidents. In many cases, each plant operator or commercial producer developed unique building designs and standard details for their own manufacturing processes. Architectural details no doubt changed or were modified as a result of lessons learned from operating experiences and as building technology changed through the years.

0097-6156/87/0345-0068$06.00/0
© 1987 American Chemical Society

Background

During early 1978 the DARCOM Project Manager's Office (DRCPM) for Munitions Production Base Modernization and Expansion Agency (currently U.S. Army Munitions Production Base Modernization Agency) requested the U.S. Army Engineer Division, Huntsville (USAEDH) to prepare standard details for use in the design and construction of buildings in which nitroglycerin, nitrocellulose, single base and multibase propellants are manufactured. The primary objective of this effort was to enhance safety and achieve standardization with possible cost reductions. A document was prepared by USAEDH entitled "Standard Details Study for NG, NC, SB & MB Facilities." This document defined the technical requirements, scope, approach, and resources required for developing the standard details. The document contained pertinent safety regulations required by the AMC Safety Manual and current practices utilized in the modernization and expansion program for facilities used in the manufacture of explosives and propellants. In addition, it outlined the proposed procedures for development and control of standard details which would be utilized in the renovation of old facilities and design of new facilities. In 1979 the DARCOM Project Managers Office authorized USAEDH to proceed with the development of the standard details. Black & Veatch was then selected by USAEDH to develop the standard details. This task was completed in December 1981 with the publication of the "Architectural Standard Details for Nitroglycerin, Nitrocellulose, Single Base and Multibase Facilities at Army Ammunition Plants," which is the basis for this paper.

For facilities susceptible to the contamination of nitroglycerin liquids and vapors, basic construction materials of wood framing, reinforced concrete, fiberglass reinforced plastic, and sandwich panels were chosen for development of architectural details incorporating lead conductive floor lining, equipment doors, personnel escape chutes and doors, ceiling and wall interfaces, interior finishes, joint sealing, door and wall louvers, wall vents, wall penetrations, and fixed windows.

For facilities susceptible to nitrocellulose, single base and multibase dusts, the same details could be used with the addition of alternate basic construction types. Six types of construction were chosen which included wood frame, concrete masonry units, reinforced concrete, modified preengineered buildings, fiberglass reinforced plastic and sandwich panels. These were chosen for development of architectural details similar to those mentioned above for nitroglycerin facilities except troweled-on conductive floor lining was to be used instead of lead.

Purpose And Objectives

The purpose of the architectural standard details is for use in the design and construction of facilities used in the manufacture, maintenance, inspection, and storage of explosive materials. To this end two objectives were sought. The requirements for this program were to develop standard details for various methods of construction utilized in Army ammunition plants today and to develop details utilizing new materials of recent development used in similar industries having the potential to increase safety, increase energy conservation, reduce maintenance and costs. The secondary objective was to establish a procedure whereby the architectural standard details can be updated to reflect "lessons learned" and to incorporate new materials and techniques as they become available.

The figures which follow represent typical nitroglycerin facility architectural details appearing in the standard details. It should be noted that these details indicate wood construction for the NG facilities which is normally not allowed by AMCR 385-100, however, these details have been reviewed and approved for use by the Department of Defense Engineering Safety Board (DDESB). In order to comply with the AMC Safety Manual, approvals may have to be obtained on an individual project basis.

It should be stressed that it is not the intention that the standard details be used directly on an ammunition plant construction project by merely specifying a particular detail by drawing number. The details should be modified to suit each particular manufacturing operation or end product and should be redrawn on contract drawings.

The following statement appears consistently on the details and will determine the choice of all materials including the basic building construction system chosen, special floor coatings, conductive flooring, interior finishes and construction sealants.

"All construction materials shall be certified to be compatible with process materials and end products. Certification tests shall be conducted on each lot of construction materials to be used in the facility."

Basic Floor Design Considerations

Basic floor design requirements that should be considered during initial design or modification of munitions production buildings are as follows:

- Surfaces should facilitate cleaning.
- Cracks and crevices where explosives particles may lodge should be omitted.
- Subfloor and finish floor surfaces chosen must not wrinkle or buckle under operating conditions.
- In chemical munitions facilities, surfaces must be sealed by coating or treating to prevent agent absorption during spills so that decontamination can be obtained.
- Porous materials should not be used for flooring.
- Coating or sealing materials must not react with agent.
- Surfaces should be capable of receiving repeated washings with hot water.
- In explosive facilities and locations where the atmosphere may contain combustible dusts, or flammable vapors or gases, ferrous metal surfaces should not be coated with aluminum paint due to the potential sparking hazard.
- Nonsparking floors are required where exposed explosives are present.
- Cove bases at the junction of walls and floors are recommended.
- Avoid exposed nails, screws or bolts.

Typical Standard Details for Wood Frame Construction

Figure 1 indicates a typical nitroglycerin facility "inside out" wood frame construction at a concrete floor slab. Note that the exterior cant strip, the lead conductive floor cant and the wood cap are all sloped to discourage product build-up and facilitate cleaning. This assembly also indicates spray-on foam insulation as an optional construction item. At Radford AAP this is a safety approved insulation system. The insulation at Radford AAP received a chlorinated rubber paint coating for weathering.

Figure 2 is a detail of the sloped wood cap used in Figure 1. Note that the joints are taped (at the top of the cant) and caulked (between the lead flooring and wood cant) to keep manufacturing components and product out of joints. The tape material is 3 inch wide, 2 ply, 100 percent cotton, grade B fabric with a warp and fill of approximately 78 x 78 x 72 pounds breaking strength. It should be adhesive-applied using a water insoluble nitrile rubber/resin solution. These are commonly referred to as "Airplane Fabric" and "Pliobond 20" adhesive. The Fiberfrax Paper is used below lead flooring as an insulation barrier with a low thermal conductivity to resist heat required for installation of lead conductive floor. Note also that nonsparking nails are required. These are usually aluminum or brass.

Basic Design Considerations for Interior Surfaces of Walls, Roofs and Ceilings

- Interior surface finishes should be -
  - Smooth.
  - Fire retardant.
  - Crack and crevice free.
  - Joints taped and sealed.
  - If painted, covered with hard gloss paint to facilitate cleaning and minimize impregnation of finish wall and ceiling materials with explosives particles.
- For horizontal ledges which might hold dust -
  - Avoid completely or bevel.
- In chemical manufacturing facilities, construct walls and ceilings of nonporous materials.
- Walls and ceilings must not absorb agent, must decontaminate easily and resist action by liquid or gaseous agents.
- In explosives buildings, roofs and walls not specifically designed for protection of personnel and equipment shall be light in weight as practicable (weak) and so constructed and supported that they will vent an internal explosion with the formation of minimum sized missiles.
- Containment structures for chemical munitions should be designed to contain both the forces of explosion and the agent dispersed by the explosion.

Figure 3 indicates a roof detail at an exterior wall. Note that the upper surfaces of joists are detailed to be sloped to minimize dust collection and that all interior joints are taped to prevent manufacturing components and product from entering joints. Exterior surfaces of insulation should receive a coating of weather-resistant paint.

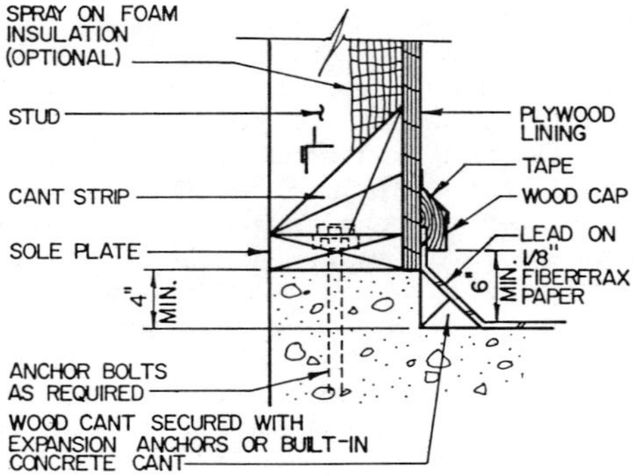

Figure 1. Exterior wall at concrete slab.

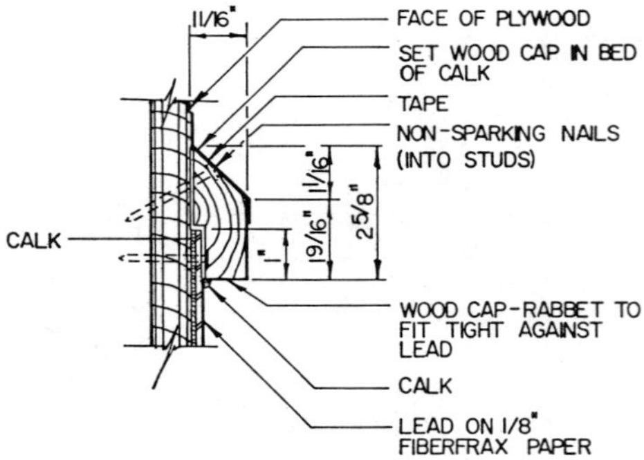

Figure 2. Wood cap detail.

Figure 4 utilizes similar sloped cant strips, taped joints, sloped conductive floor cant and wood cap above lead flooring. Note the use of non-sparking aluminum or stainless steel exterior flashing.

Basic Design Considerations for Building Exits and Windows

- Exit doors:
  - Should open outward in the direction of emergency egress.
  - Should not be fastened with locks other than antipanic catches or other quick releasing devices.
  - Should be casement type, glazed with nonshatterable plastic material.
  - Minimum opening size: 30 inches wide by 80 inches high.
- Windows:
  - Overall size of windows should be kept to a minimum.
  - Shatter-resistant plastic glazing should be used.

Figures 5 and 6 are details for a wood equipment door. Here again all joints are taped and surfaces of the head are canted. Note that glazing is acrylic plastic to comply with the AMC Safety Manual requirements. Screws for attachment of wood door frames and vision panel stops are countersunk and caulked. Joints not taped are sealed with caulking. It should be noted here that Sunflower AAP has had major problems with exterior wood doors exposed to the weather. A recurring problem has been the delamination of wood door materials. This may require a change to a more weather-resistant door material such as fiberglass reinforced plastic. Details for doors of this material are included in the standard architectural details.

Figure 7 is a window detail indicating positioning, for safety reasons, of an exterior mounted light fixture for lighting the building interior. Exterior and interior of window sills are canted, including the interior trim. All joints are taped. All sparkproof metal fasteners are countersunk and caulked. The light fixture would be bracketed off the exterior window jambs.

Hardware Considerations

- In buildings containing exposed explosive materials, dusts, or vapors, hardware should be nonsparking material.
- Fasteners such as nuts and bolts which are located so tnat accidental entry into explosives or explosive constituents is possible should be securely held in place by being drilled and thonged or otherwise secured.

This series of figures, Figures 8 through 13, indicates a typical arrangement of a personnel escape door. The door is held in place by a wood pin and a nonsparking bronze or stainless steel spring catch. In Figure 9, note that this door is detailed around a fiberglass reinforced plastic material. Figure 10 indicates the standard method for securing escape doors by use of a break away hardwood latch bar. Note that the latch bar is grooved in the center near the door meeting stiles to permit rapid escape by breaking the latch bar when either or both door leaves is pushed out.

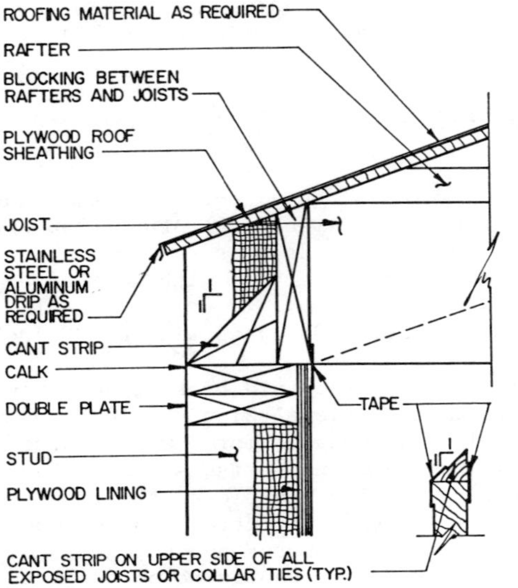

Figure 3. Exterior wall at roof.

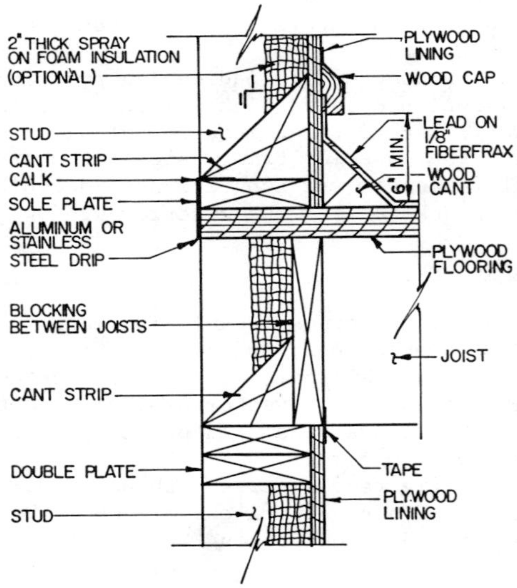

Figure 4. Exterior wall at second floor.

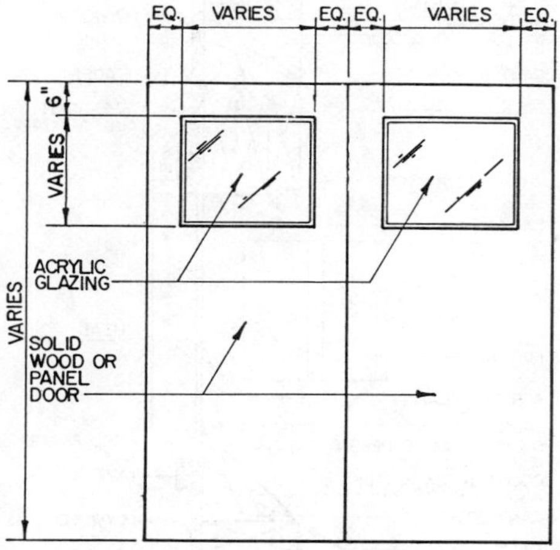

Figure 5. Elevation of wood equipment door.

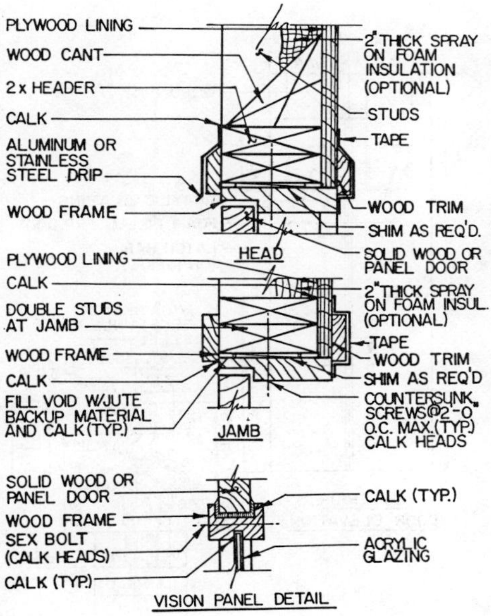

Figure 6. Door details.

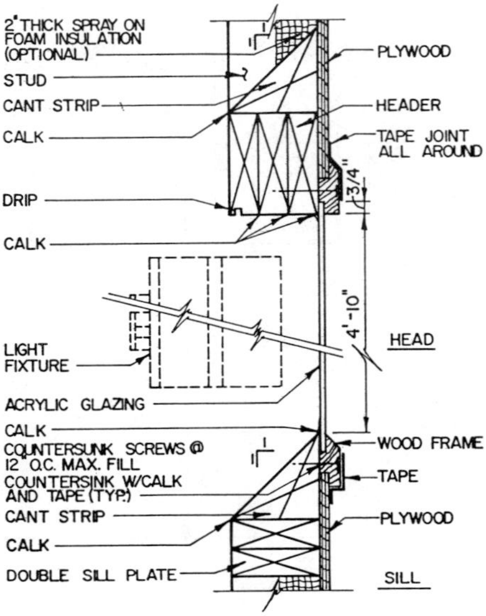

Figure 7. Exterior lighting/window details.

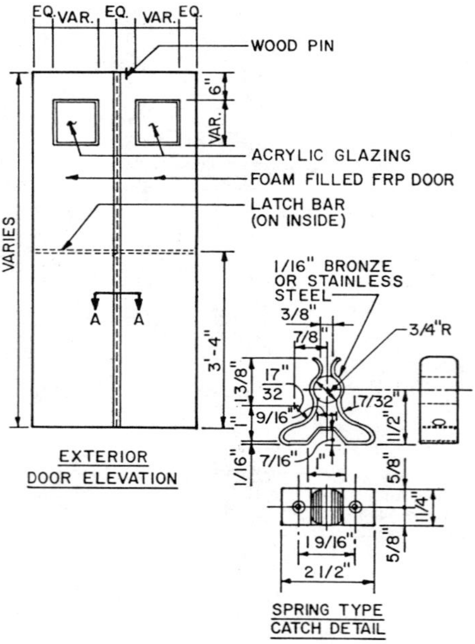

Figure 8. FRP personnel escape door.

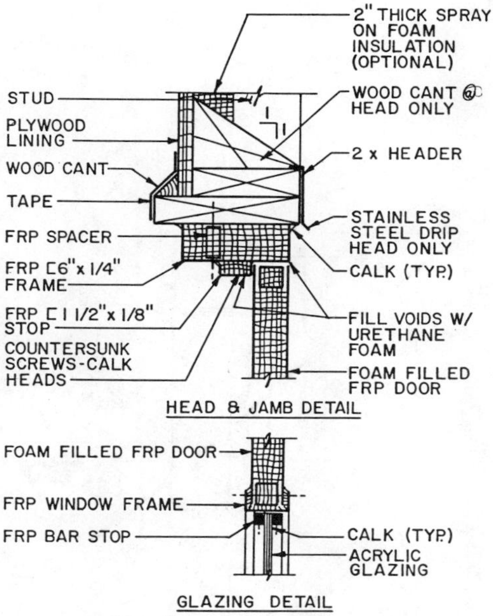

Figure 9. FRP personnel escape door details.

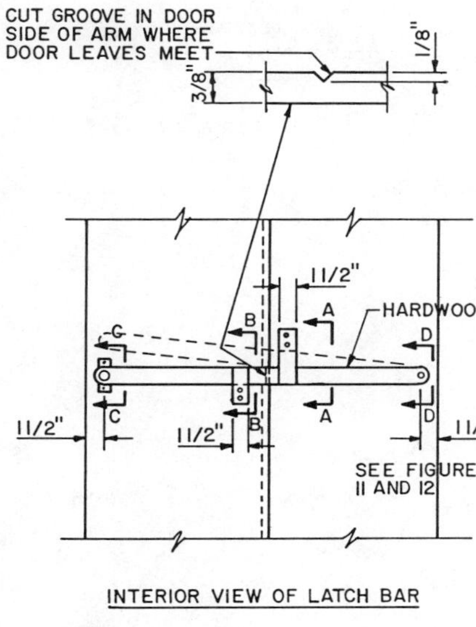

Figure 10. Door latch bar details.

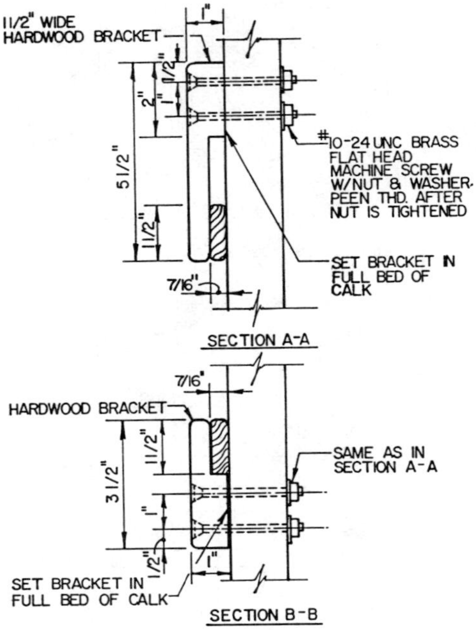

Figure 11. Door latch bar details.

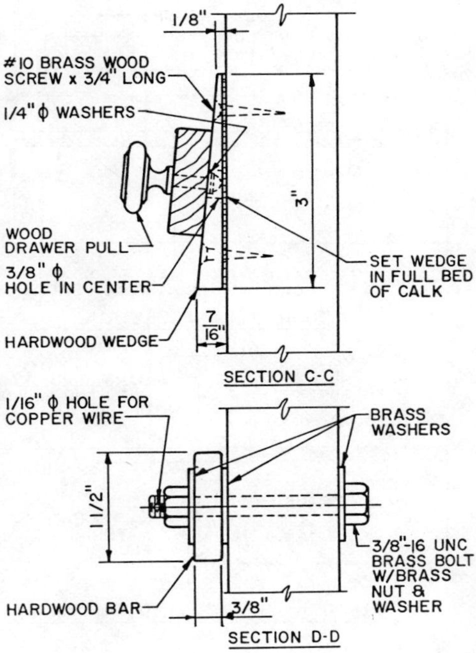

Figure 12. Door latch bar details.

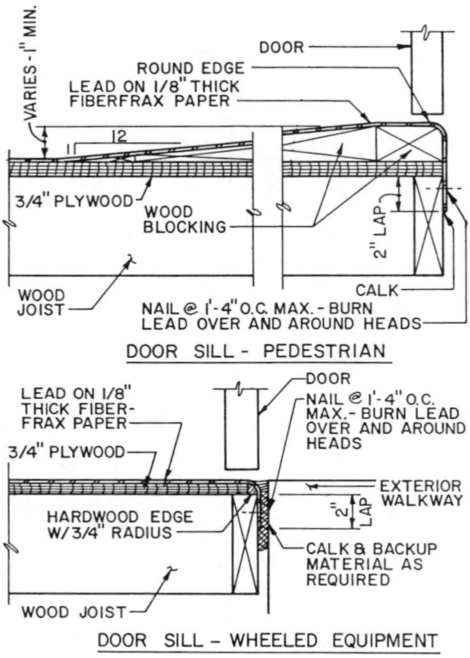

Figure 13. Door sill details.

Figures 11 and 12 detail the latch bar. Note that nonsparking metal is used for all fasteners and that the hardwood wedge in Section C-C is set in a full bed of caulking so as not to permit an open joint.

Figure 13 represents two door sill conditions. The pedestrian door sill is required at locations where the product is not permitted to drain out to the exterior. Note that 1 in 12 slope towards the interior.

The door sill for wheeled equipment is a flat sill meeting the entrance pavement elevation providing a level transition in or out.

A typical interior trench or floor gutter is shown in Figure 14. Note the rounded bottom shape and the canted or rounded bends of the lead conductive flooring. Also, note the requirement for rounding bearing surfaces of the cover, which prevents damage to the lead floor surfacing.

Floor Gutter Design Considerations

- Gutters should be free of pockets.
- Sufficient slope is required (1/4 inch per foot minimum).
- Gutters inside buildings may be sloped 1/8 inch per foot minimum.
- Drains between the source of explosive and sumps shall be troughs with rounded bottoms and ventilated covers to facilitate inspection for accumulation of explosives.

Figure 15 is a new standard design for a fiberglass reinforced plastic (FRP) escape chute which replaces existing sheet metal escape chutes. This design was based on a standard detail furnished by the Corps of Engineers, Huntsville Division.

Note that the chutes are fabricated of standard FRP sections with reinforcing rib members. Sections are bolted together. Note also that an integral support column is necessary for safety chutes extending above a second floor. The radius of the chute is shown as 6'-0" at the bottom.

Safety Chute Design Considerations

- Exits to safety chutes should open onto platforms not less than 3 feet square that are equipped with guardrails.
- Safety chutes should begin at the outside edge of platforms.
- Recommended safety chute specifications are as follows:

    Slope angle: 40° to 50° with horizontal

    Chute depth: 24 inches

    Radius at bottom of chute: 12 inches

    One additional foot of horizontal run should be provided for each additional 5 feet of chute length.

Procedure for Making Changes

Advances in technology, architectural/engineering practices or advances gained from the experience from the actual on-site performance of certain standard details installed at Army ammunition plants will naturally lead to proposed changes and additions or deletions from the baselined standard details. These changes will not be discouraged. The procedure for making proposed changes as stated in the Architectural Standard Details is as follows:

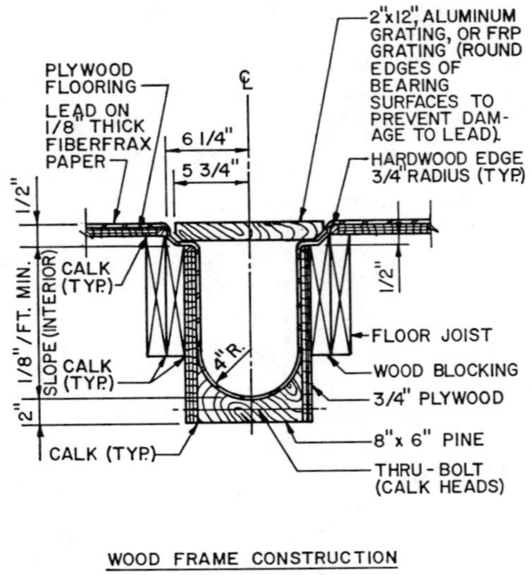

Figure 14. Lead conductive floor. Floor gutter/floor interface.

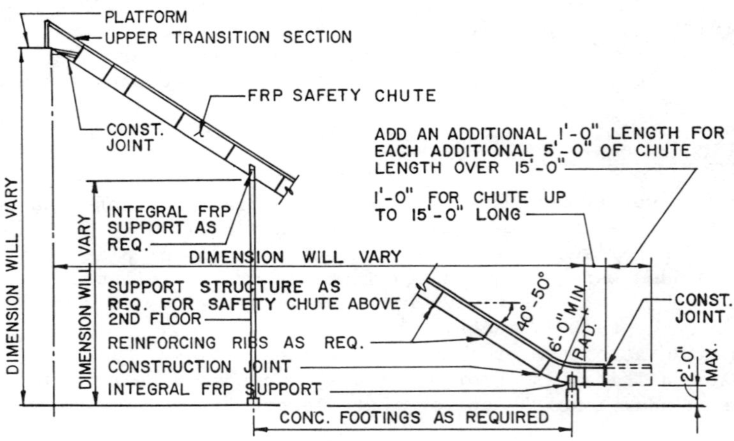

Figure 15. FRP escape chute.

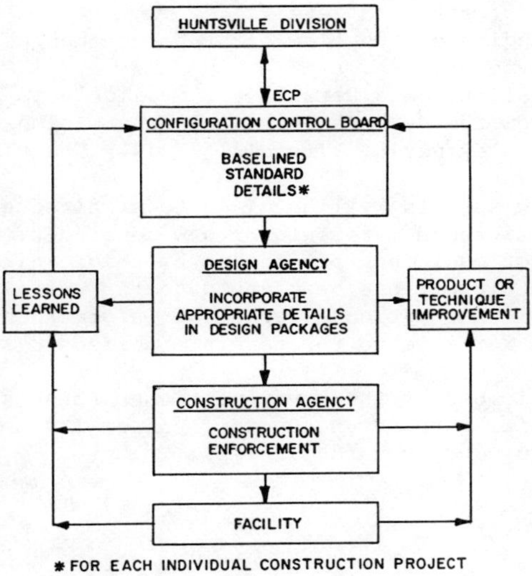

Figure 16. Flow of proposed changes during the review and approval process.

1. Proposed changes, additions or deletions regardless of their originating agencies or the nature or purpose of the change must be processed as an Engineering Change Proposal (ECP).
2. The changes are then reviewed by the various concerned agencies.
3. Final approval will then be made by the Configuration Control Board (CCB).
4. The Corps of Engineers, Huntsville Division will serve as the focal point for coordinating all activities associated with the modification of standard details.

Figure 16 indicates the flow of proposed changes during the review and approval process.

Architectural Standard Details are available to anyone who requests them from the Defense Technical Information Center (DTIC), Cameron Station, Alexandria, Virginia 22314. The DTIC acquisition number is AD-A112 677.

The standard details will be given to architects and engineers as criteria or reference material for new construction or modification design for munitions production base modernization.

The document itself has been approved for unlimited distribution and is included in the National Technical Information Service (NTIS) listings. It is for sale to the general public and foreign nationals.

It is anticipated by the Government that these standard details will serve a useful purpose in assuring uniformity and safety in future AAP designs for such facilities.

Acknowledgments

Acknowledgement is made of the cooperation extended by Mr. Richard Mazinski, U.S. Army Munitions Production Base Modernization Agency, Department of the Army, and Mr. Morgan F. Jones, formerly with the U.S. Army Engineer Division, Huntsville. Their cooperation in furnishing information essential to this paper is appreciated.

Literature Cited

1. Standard Details for NG, NC, SB, and MB Facilities, U.S. Army Corps of Engineers, Huntsville Division, HNDTR-78-34-ED-SR, Revision 1, 15 January 1979.
2. Architectural Standard Details for Nitroglycerin, Nitrocellulose, Single Base & Multibase Facilities at Army Ammunition Plants, U.S. Army Corps of Engineers, Huntsville Division, HNDSP-81-025-ED-FD, December 1981.
3. AMC Safety Manual, AMC-R 385-100, 1 August 1985, Headquarters U.S. Army Development & Readiness Command.

RECEIVED March 6, 1987

Chapter 4

Explosives Storage Structures

Richard L. Wight

U.S. Army Corps of Engineers, Attn: DAEN-ECE-T, 20 Massachusetts Avenue, N.W., Washington, DC 20314-1000

A magazine is a unique structure with special features. Explosive contents are a threat to the magazine's vicinity, and the explosive contents themselves can face various threats. The threat factors led directly into design and construction criteria. The Department of Defense (DOD) uses standardized magazine designs which results in several benefits.

A typical explosive storage structure, or magazine, appears to be nothing more than an enlarged storm-cellar from a Midwestern farm. Actually, it is unique type of structure that incorporates special design considerations. Such a structure will be discussed in this chapter.

Function of Structure

The fundamental purpose of any storage structure is to preserve its contents until needed. Thus the purpose of a magazine is to preserve explosive material until needed. The explosive contents must be kept safe, secure, accessible, and usable. In addition, these contents must be a minimal threat to the magazine's vicinity.

Threats to Structure

Threats to the magazine's explosive contents also exist, such things as:

- -theft,
- -lightning,
- -penetration of structure by projectile,
- -fire,
- -vandalism,
- -explosive forces from accident in neighboring magazine,
- -corrosion,
- -water damage,
- -jarring or tumbling of storage containers,
- -rodents,
- -and environmental deterioration.

This chapter not subject to U.S. copyright
Published 1987 American Chemical Society

Design Requirements

The above paragraphs suggest that the ideal magazine should:

- -prevent projectile penetration,
- -prevent human (and animal) intrusion,
- -resist blast forces from outside the magazine,
- -directionalize forces from an explosion within the structure,
- -be fireproof and lightning-proof,
- -be watertight,
- -provide a stable environment,
- -provide adequate storage volume in an efficient arrangement,
- -and accommodate materials-handling equipment.

Construction Features

When the requirements of the ideal magazine are translated into construction terms, the following features could emerge:

- -Earth cover.
- -Structural shell, arch or box.
- -Concrete headwall and large steel door.
- -Screened or impassable vents and intrusion-detection system.
- -Siting so greatest distance to a neighboring magazine is away from the weakest side, the headwall.
- -Noncombustible construction and a lightning protection system with aerials and counterpoise.
- -Relatively constant temperature and humidity resulting from the earth cover.
- -Waterproofed surface and drains within the earth cover.
- -Capability of being built to various lengths.

Figure 1 is a sketch of a typical arch magazine that shows some of the features just listed.

Standardization

The DOD owns thousands of explosives storage structures that have been constructed over the past years. Many are similar to Figure 1 in concept. Today this similarity has been institutionalized and is called standardization. Standardization has several benefits:

-Safety is known. Most DOD standard designs have been "proof-tested" by exposure of a test structure to the explosive effects of a nearby detonation. The worst-case test condition is depicted in Figure 2.

-Time is saved. DOD construction projects involving explosives require special review by the DOD Explosives Safety Board. Standard designs are pre-approved; this saves review time.

-Design costs are saved. All the designer has to do is adapt the foundation to the site conditions. The rest of the design need not be touched.

-Construction cost are saved. Most explosives storage projects involve several magazines. Repetitive construction of identical structures is cost effective.

-Security is known. The weak points of standard structures against intrusions have been studied, and corrective measures have

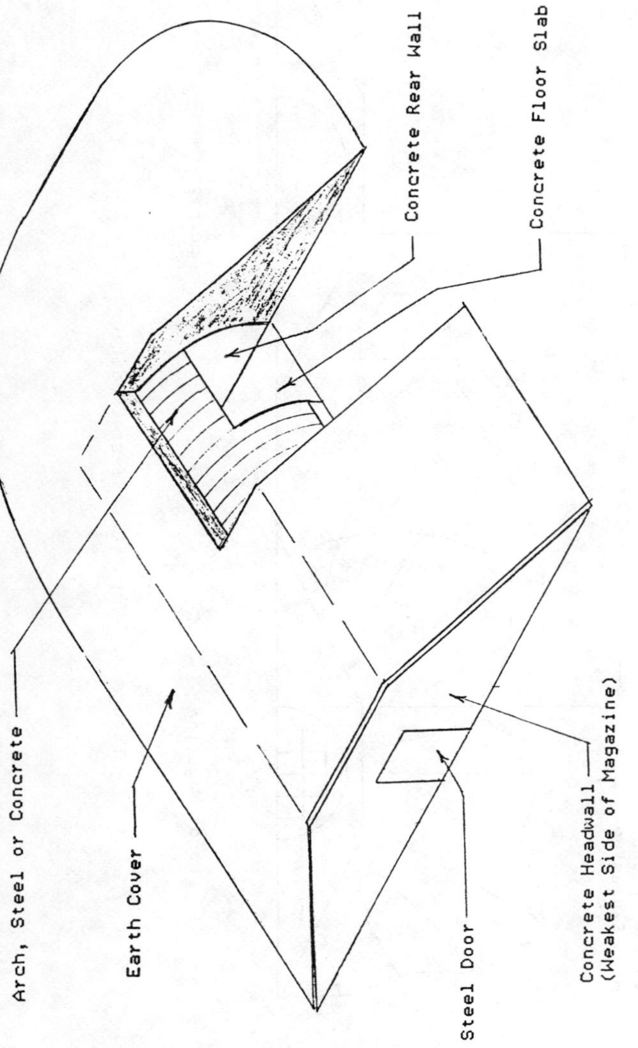

Figure 1. Structural Features of Typical Magazine

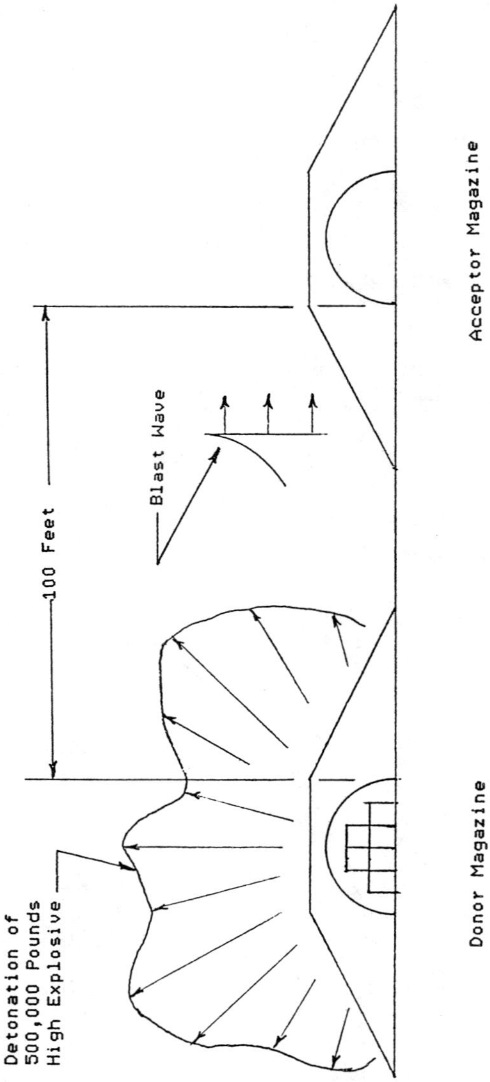

Figure 2. Worst Case Test Condition

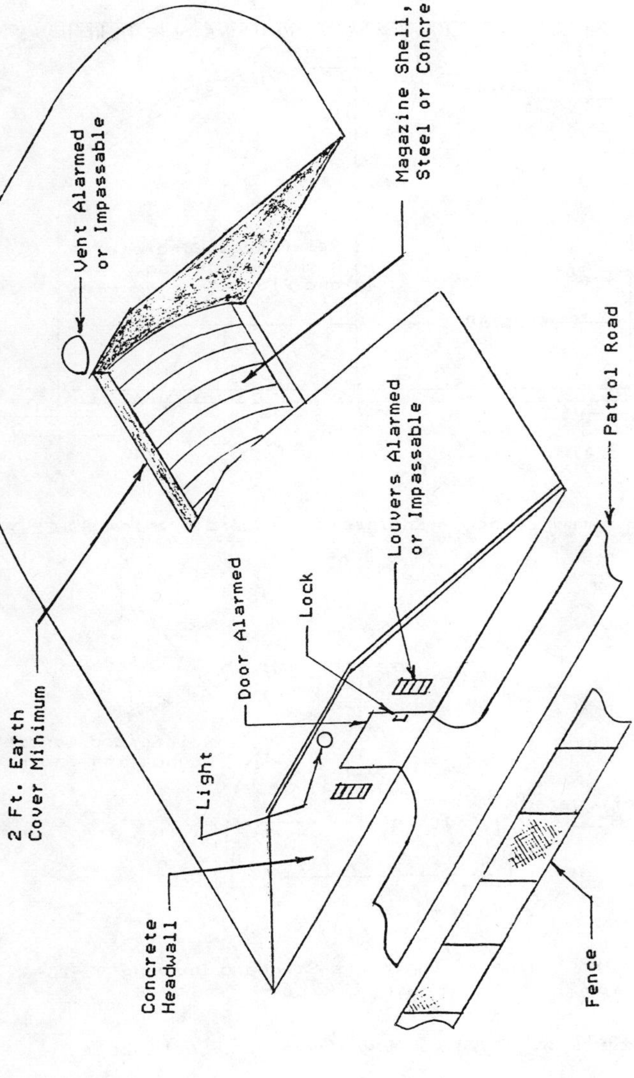

Figure 3. Security Features of Magazine Site

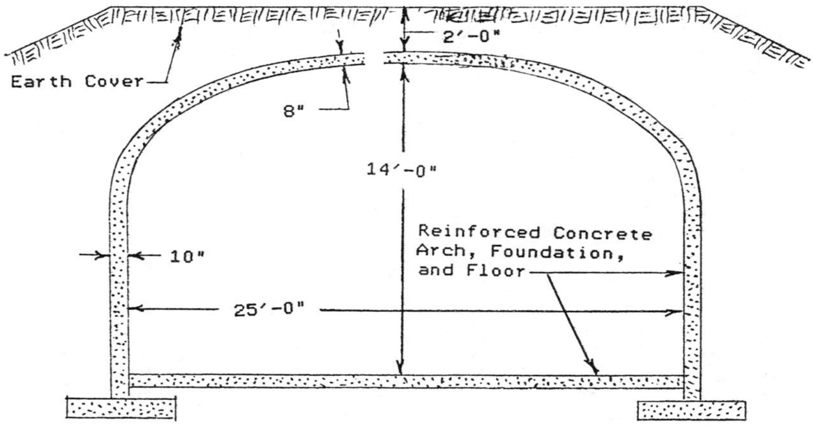

From U. S. Army Corps of Engineers Standard Drawing 33-15-74

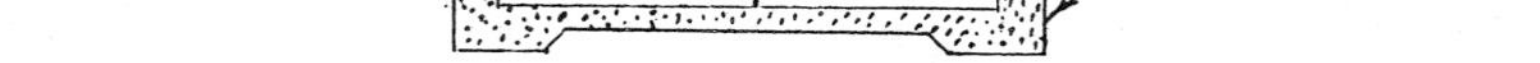

From U. S. Army Corps of Engineers Standard Drawing 33-15-65

Figure 4. Typical Magazine Cross-Sections

been developed. Figure 3 shows security-related features of a typical earth-covered magazine.

Figure 4 shows the cross-sections of two magazines designed by the U.S. Army Corps of Engineers. Both are standard magazines.

RECEIVED March 26, 1987

Chapter 5

Reinforced Concrete in Blast-Hardened Structures

James E. Tancreto

Naval Civil Engineering Laboratory, Port Hueneme, CA 93043

The design criteria for reinforced concrete is being revised as the result of dynamic tests that show improved response to explosive loads. The improved design criteria will result in safer and less expensive protective shelters and barriers. Improvements in the design criteria are mainly the result of increases in the allowable design stresses and allowable ultimate flexural deflections under shock loads. Conventionally reinforced concrete, with the proper design considerations, may now be designed for up to four times the deflections (and energy absorbing capacity) allowed by the old criteria. The improved response criteria for conventional reinforced concrete will reduce the need for more expensive laced reinforced concrete. A summary of the new design criteria is presented with emphasis on the important changes to the flexural design criteria.

Explosive storage and operating facilities must be designed to protect personnel, equipment, and contents from the effects of an accidental explosion. Hardened structures can be classified as shelters or barriers. Shelters are designed to completely shelter their contents from the blast and fragments produced by an explosion. Barriers are walls or open structures that provide partial protection. Barriers are usually designed to prevent sympathetic detonation of explosives by stopping fragments and reducing blast pressures from an adjacent explosion.

Reinforced concrete is the most commonly used construction material for structures designed to resist explosive blast loads. It is used extensively in blast hardened structures because of its strength, ductility (when properly designed), mass, penetration resistance, relative economy, and universal availability. Its strength, mass, and ductility provide high resistance to the extreme blast pressure (psi) and impulse (psi-ms) loads. It is important to remember that (unlike in static load design) in the

This chapter not subject to U.S. copyright
Published 1987 American Chemical Society

design for dynamic loads, the mass and ductility of the element are as important as its strength. The mass and strength also provide excellent fragment and debris penetration resistance. Unhardened reinforced concrete, designed for normal (non-explosive) loads will generally be much more blast resistant than other structural materials because of these attributes.

Blast hardened reinforced concrete structures may still be very massive and expensive. The expense increases when lacing reinforcement is necessary to provide ductility at the large deflctions caused by severe blast loads. Tests have shown that conventional reinforced concrete (without lacing) can attain much larger deflections, with proper design, than are being allowed by existing criteria. New criteria are being developed to reflect these test results. A summary of the new evolving criteria, especially the bending criteria as reflected in the tri-service design manual, TM 5-1300/NAVFAC P-397/AFM 88-22, "Structures to Resist the Effects of Accidental Explosions", is presented here.

Behavior Modes

Two modes of behavior, ductile and brittle, must be considered in the design of hardened reinforced concrete structures. Reinforced con crete can behave with great ductility during the flexural response of bending members (slabs, beams, girders, etc.). This ductile flexural mode results in large deflections that can absorb the high energy from the blast loads. The brittle modes (shear failure, compression failure, spalling, breeching, and fragment penetration) may reach failure under relatively low energy input levels or at small deflections due to load concentrations and low ductility. Brittle failures occur before significant bending deflection can develop. Reinforced concrete bending elements are designed to resist the blast loads in the high energy absorbing flexural mode and then shear reinforcement is provided to prevent an early shear failure. A basic design requirement for reinforced concrete is that flexural elements be designed so that failure is forced to occur in bending and not shear.

<u>Ductile Behavior</u>. When a reinforced concrete element is loaded by the blast load it deflects elastically until plastic yielding occurs along highly stressed yield lines. It then deflects plastically (with a small increase in resistance from strain hardening of the steel) to its maximum deflection. Figure 1 shows a typical resistance deflection curve. The degree of ductility is represented by the maximum support rotation (and center deflection) that can be attained without failure. Figure 2 shows the relationship between support rotation and maximum deflection of a one-way bending member. The relationship for a one-way element is:

$$X = (L/2) \tan \theta$$

where X = deflection (at center span of one-way member)
θ = angle of rotation at support

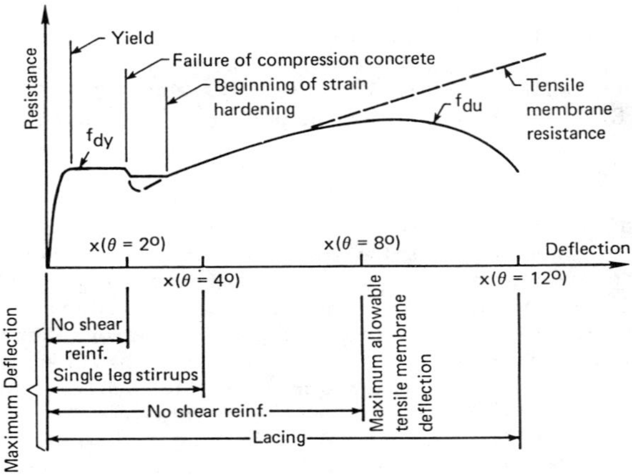

Figure 1. Typical resistance-deflection curve for flexural response of concrete elements.

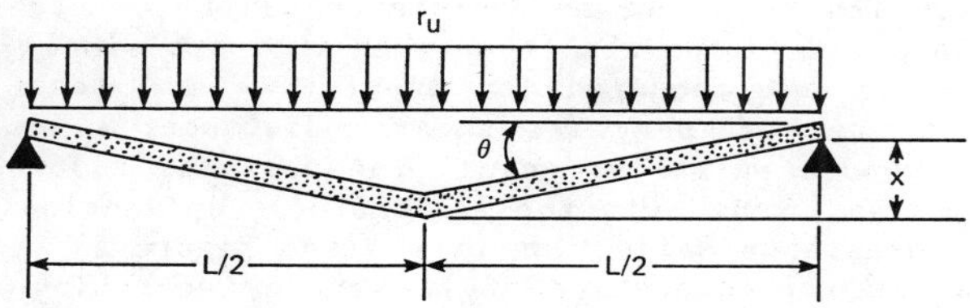

Figure 2. Deflection of a one-way simply supported bending element.

The relationship for two-way elements (such as slabs supported on three or four sides) is more complicated. An approximate relationship can be obtained for any element by substituting the short span for L/2 in the above equation. The allowable rotation and deflection is strongly dependent on compression and buckling strength of the reinforcement on the compression side of the element. Previous criteria have allowed a design support rotation, θ, of 2 degrees for conventionally reinforced concrete. When lacing steel (as shown in Figure 3a) is used to prevent buckling of the compression reinforcement and to contain the cracked concrete, a design support rotation of 12 degrees is allowed.

Testing has shown that conventionally reinforced concrete (without lacing) can safely sustain much larger support rotations than 2 degrees. The new criteria are taking advantage of these test results to allow increased support rotations and center deflections. The changes include allowable support rotations of 4 degrees for conventionally reinforced concrete (with single leg stirrups, as shown in Figure 3b, to increase ductility) and 8 degrees for reinforced concrete that can develop tensile membrane resistance. Tensile membrane resistance can be counted on in most two-way slabs and flat slabs (even when they are simply supported). Shear steel is not required for ductility in a tensile membrane slab but may be necessary for shear resistance. These increased allowable support rotations result in increased allowable deflections of two and four times the old criteria deflections. The area under the resistance deflection curve (see Figure 1) between $X = 0$ and $X = X_u$ is representative of the energy absorbing capacity of the structure. Thus, increasing the allowable design deflection proportionally increases the area under the resistance-deflection curve. Figure 4 shows the design elasto-plastic and perfectly plastic (for support rotations > 5 degrees) resistance-deflection functions.

The increased impulse capacity of a structure is proportional to the square root of the increase in the area under the resistance-deflection curve. The effect of mass can be easily shown with the following equation for the impulse capacity of a ductile element with large allowable deflection and a perfectly plastic resistance function (as shown in Figure 4b).

$$i^2 = 2\, m_u\, r_u\, X_m$$

where i = blast load impulse, psi-ms

m_u = effective unit mass in ultimate range, psi-ms/lb$^{1/3}$

r_u = ultimate unit resistance, psi

X_m = maximum deflection, in

In the equation above, mass carries the same "weight" as strength and ductility (deflection) in developing impulse capacity.

The allowable support rotation and deflection for laced reinforced concrete has remained at 12 degrees. The increased allowable deflections for conventionally reinforced concrete will reduce

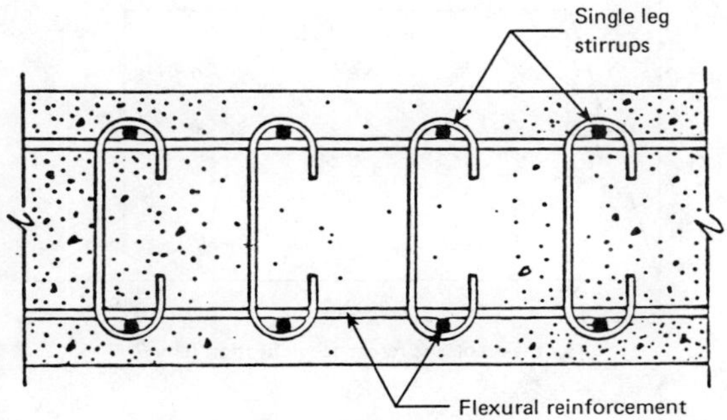

b. Single leg stirrups.

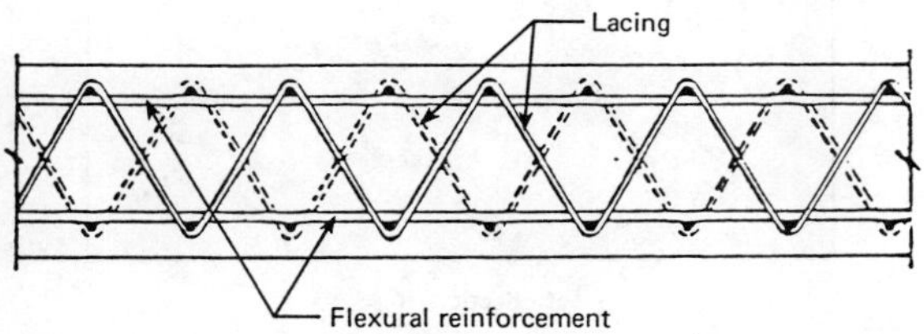

A. Lacing reinforcement.

Figure 3. Typical shear reinforcement.

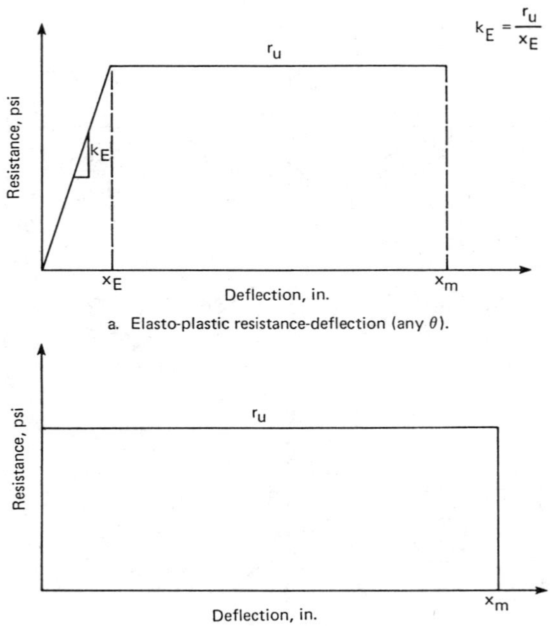

a. Elasto-plastic resistance-deflection (any θ).

b. Perfect plastic resistance-deflection ($\theta > 5^{o}$).

Figure 4. Typical design resistance-deflection functions.

the need for expensive laced reinforced concrete in hardened structures.

Brittle Behavior. Three related brittle modes of failure create concrete fragments during bending response: spalling, scabbing and post-failure fragmentation. Spalling and scabbing consist of concrete debris from the concrete cover over the flexural reinforcement. Spalling occurs before significant bending can begin and is caused by high tensile forces created by the blast pressures. Scabbing, a form of spalling, occurs at large bending deflections when severe cracking of the concrete cover has occured. Post-failure concrete debris are created from the collapse of an element and are usually numerous, large and have relatively high velocities. Spalling and scabbing can be hazardous to personnel, sensitive equipment, and sensitive explosives. Spalling and scabbing can be controlled with spall plates, and by limiting design deflections. Postfailure fragments are avoided by designing to prevent failure (the normal hardened structure design requirement).

Other brittle failure modes include shear (direct and diagonal tension), compression failure, breeching, and fragment penetration. Bending elements must be designed to develop their full bending capacity. Shear failures are controlled by providing reinforcement adequate to support the full bending resistance (r_u) of the member. Compression failure is controlled with proper distribution of the reinforcement (usually equal steel percentages on the tension and compression sides) and, for design rotations above 2 degrees, lateral support of the compression reinforcement with single leg stirrups or lacing. Underreinforced sections are used in design to keep the shear and compression stresses low, allowing ductile bending response to develop before shear or compression failure can occur. Axial compression members (columns) are designed to provide adequate compression and shear strength to support the ultimate resistance of supported bending members.

Breeching is a local perforation of the concrete element by the extremely high blast pressures of a close explosion. High velocity concrete fragments can result. Breeching failures are controlled by providing adequate reinforcement, concrete thickness and standoff distance to the explosive.

Reinforced concrete is very resistant to fragment penetration and is frequently used just for this reason. Primary fragments can produce spalling of the concrete. Perforation by metal fragments and concrete spalling are controlled by providing adequate concrete thickness based on empirical relationships using fragment mass and velocity.

Dynamic Strength of Materials

The allowable strength of materials is higher under dynamic loads, which produce high strain rates, than under static loads. This results in higher resistance to dynamic loads. The most important increases are in the compression strength of concrete and the yield strength of the steel reinforcement.

Static Strength. ASTM A 615, Grade 60 reinforcement, is recommended for hardened reinforced concrete design. The average yield strength for this steel is 10 percent greater than the minimum required ASTM value (60,000 psi), while the ultimate strength is not much greater than the ASTM minimum. The recommended static yield and static ultimate design strengths are:

$$f_y = 66{,}000 \text{ psi} \quad \text{and} \quad f_u = 90{,}000 \text{ psi}$$

In the design calculations for flexural elements, the concrete strength is only important in determining the shear resistance of elements undergoing less than 2 degrees support rotation. However, stronger concrete will also result in less cracking and crushing of concrete between the reinforcement at large rotations. It is recommended that the design concrete static design strength be 4000 psi, and never less than 3000 psi.

Dynamic Strength. The dynamic design strengths for steel reinforcing and concrete are equal to their static design strengths times the appropriate Dynamic Increase Factor (DIF).

$$f_{(dynamic)} = DIF \times f_{(static)}$$

Table I summarizes the appropriate DIF's by type of stress.

Table I. Dynamic Increase Factors (DIF) for Reinforced Concrete

Type of Stress	Low-Intermediate (& High) Design Pressures*		
	Reinforcing Steel Yield	Reinforcing Steel Ultimate	Concrete Ultimate
Bending	1.17 (1.23)	1.05	1.19 (1.25)
Diag. Tension	1.00 (1.10)	----	1.00
Direct Shear	1.10	1.00	1.10
Bond	1.17 (1.23)	1.05	1.00
Compression	1.10 (1.13)	----	1.12 (1.16)

*The revised Tri-Service design manual uses Far and Close-in Design Ranges rather than Low-Intermediate and High Design Pressures

Flexural Design

Flexural member design requires the determination of: (1) the design blast loads, (2) the initial design cross-section, (3) an idealized resistance deflection function, (4) the calculated response (maximum deflection) and, (5) allowable ultimate deflection and (6) design for shear.

Blast Loads. The flexural member is designed for the expected blast overpressure loads (pressure and impulse). External blast overpressure loads are primarily dependent on the equivalent explosive weight (W), the range from the structure (R), and the orientation of the structure to the shock wave. The design loads include the effect of a 20% factor of safety on the explosive weight (Design Explosive Weight = 1.2W). Other factors, including charge shape, the height of burst (HOB), terrain effects, and casing thickness, can influence the blast overpressure and impulse loads and are included in the loads determination when possible.

The loads from external near-surface burst explosions are based on hemispherical surface burst relationships. Peak pressure (P psi) and scaled impulse ($i/W^{1/3}$ psi/lb$^{1/3}$) are plotted vs. scaled distance ($R/W^{1/3}$ ft/lb$^{1/3}$). Roof and sidewall elements, side-on to the shock wave, see side-on loads (P_s and i_s). The front wall, perpendicular to the shock wave, sees the much higher reflected shock wave loads (P_r and i_r). An approximate triangular pressure-time relationship is shown in Figure 5a. The duration, T, is determined from the peak pressure and impulse by assuming a triangular load. Complete load calculations include dynamic loads on side-on elements, the effect of clearing times on reflected pressure durations, and load variations on structural elements due to their size and varying distance from the explosive source.

Internal explosive loads include direct reflected shock pressures plus (1) the reflected shock pressures from adjacent surfaces and (2) internal gas pressures from the gaseous products of the explosion. The peak gas pressure, which is a function of the charge density (charge weight to structure volume ratio, W/V), is relatively low but can be of long duration with large impulse. Frangible surfaces are commonly used to quickly vent the gas pressures and reduce the internal design load on the hardened structure. The direct plus reflected internal shock pressures and the gas pressures can be determined from curves in NAVFAC P-397. A bilinear load function is obtained by merging the shock pressure and gas pressure curves as shown in Figure 5b.

Trial Cross-Section. A trial cross-section is chosen that includes the concrete thickness, and the tension and compression steel percentages (in the horizontal and vertical directions for a two-way slab). The optimum distribution of horizontal and vertical steel is obtained when 45 degree yield lines are obtained in a yield-line analysis for ultimate resistance. The minimum steel percentage, either way and in tension or compression, is 0.15%. The optimum total positive or negative reinforcement ratio ($p_h + p_v$) has been found to be between 0.6% and 0.8%. A value in this range should be used for design.

Resistance-Deflection Function. The resistance-deflection function establishes the dynamic resistance of the trial cross-section. Figure 4a shows a typical design resistance-deflection function with elastic stiffness, K_E (psi/in), elastic deflection limit, X_E (in) and ultimate resistance, r_U (psi). The stiffness is determined from a static elastic analysis using the average moment of inertia of a cracked and uncracked cross-section. (For design

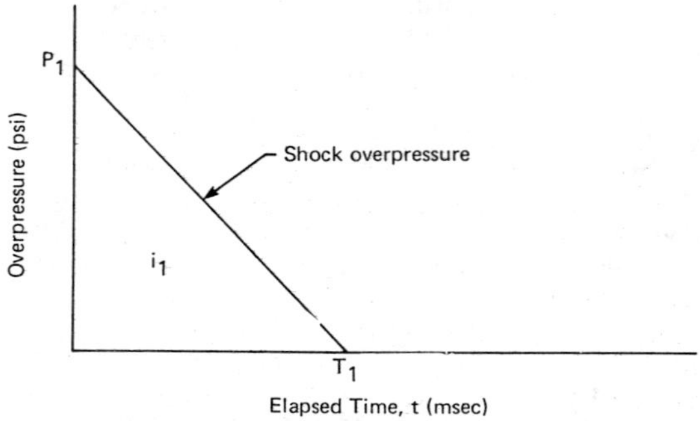

a. Typical external design load function.

P_1 = peak shock overpressure, psi
P_2 = peak gas overpressure, psi
T_1 = duration of design shock load, msec
T_2 = duration of design gas load, msec
i_1 = shock inpulse, psi-msec
i_2 = gas impulse, psi-msec

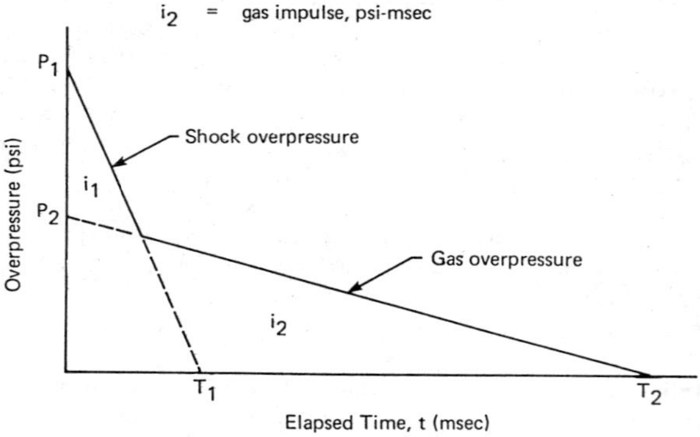

b. Typical internal design load function.

Figure 5. Design overpressure versus time.

deflections > 5 degrees, the perfect-plastic resistance-deflection function in Figure 4b may be used. This eliminates the need for determining the stiffness of the element.) Ultimate resistance is determined statically using yield-line analysis and dynamic stress allowables. The ultimate resistance is the uniform static pressure that the element can support when yielding begins at X_E. The ultimate resistance depends on the moment resistance of the cross-section, the geometry of the element, and the support conditions. The moment resistance of the section changes with increasing deflection as the concrete cover crushes (reducing the moment capacity slightly) and as the steel reinforcement strain-hardens and increases in strength from yield to ultimate (increasing moment capacity). These variations are averaged, depending on the design deflection, to obtain the straight line design resistance functions shown in Figure 4.

Maximum Deflection. The response of the trial section is determined from an equivalent single-degree-of-freedom (SDOF) spring-mass system. Response charts are available for the triangular or bilinear load functions (see Figure 5) and an elastic plastic resistance function (see Figure 4). The response charts give X_m/X_E versus P/r_u and T/T_N. B and T are the peak pressure and duration, respectively, in the load function. T_N is the natural period of the equivalent SDOF spring-mass system. The natural period is given by:

$$T_N = 2\pi(m_e/K_E)^{1/2}$$

where $m_e = mK_{LM}$

m = unit mass of the element (psi-ms/lb$^{1/3}$)

K_{LM} = SDOF load-mass factor

The load-mass factor, K_{LM}, transforms the actual dynamic system to the equivalent SDOF system. The value is usually between 2/3 and 3/4 and depends on the geometry, end conditions, support conditions, and range of behavior (i.e. elastic, elasto-plastic, or plastic). The maximum deflection, X_m, is then compared to the allowable ultimate deflection to determine the adequacy of the trial section.

Allowable Deflection. The allowable deflection is directly calculated from the allowable support rotation and the shortest distance from a support to a yield-line (L/2 for a one-way element). The allowable support rotation depends on the ductility of the section as summarized in Table II.

Tensile membrane behavior requires continuous reinforcement steel to support in-plane stesses. Two-way slabs and flat slabs, with fixed or simple supports, can usually satisfy the requirements for tensile membrane resistance. Design with tensile membrane resistance is the same as for flexural resistance since the moment capacity of the section is used to determine ultimate resistance. Tensile membrane resistance at 8 degree rotation must be at least equal to the bending resistance to insure that adequate strength is available when bending resistance is lost.

Table II. Allowable Support Rotations, θ_u

Shear Reinforcement	Tensile Membrane Resistance	θ_u* (deg)
None**	No	2
Stirrups	No	4
None**	Yes	8
Lacing	No	12

*Does not apply if containment is required.
**Not required for ductility but must be used if required for shear.

If the maximum deflection calculated for the trial section is less than the allowable deflection, then the section is adequate in bending and the shear stresses must be checked.

Shear Design.

The shear loads, V_u, are based on the ultimate bending resistance, r_u, of the structural element. Shear resistance is provided to support the resulting shear stresses, v_u. This allows the element to reach its full dynamic flexural load carrying capacity and not fail prematurely, in shear, at small deflection. Two major shear stresses must be checked: diagonal tension at a distance from the support, and direct shear at the support.

Diagonal Tension. The allowable shear stress, v_c (psi), on a concrete section without shear reinforcement is:

$$v_c = 1.9(f'_{dc})^{1/2} + 2500\ p < 3.5(f'_{dc})^{1/2}$$

where f'_{dc} = allowable dynamic concrete compression stress, psi

p = tension reinforcement ratio

When the shear stresses exceed the allowable for an unreinforced section, then shear steel must always be used to provide the additional strength (to take excess shear $v_u - v_c$). In addition, when stirrups or lacing are used for obtaining allowable support rotations of 4 degrees or 12 degrees respectively (see Table 2), then the shear steel must be designed for a minimum excess stress of $0.85v_c$.

Direct Shear. For type I cross-sections ($\theta < 2^o$) the concrete between the flexural reinforcement is capable of resisting direct shear. However, because cracking at the support yield line reduces the shear capacity, diagonal bars must be provided to at least resist the shear capacity of the concrete, v_c. For type II and III cross-sections ($\theta > 2^o$), with little or no concrete shear resistance, The diagonal reinforcing bars must be designed to resist the entire shear load at the support.

Design for Spalling, Breeching, and Fragment Penetration. Test results have been used to empirically derive relationships for the reinforced concrete thickness required to prevent spalling, breeching, or fragment penetration.

Design Criteria for Breeching and Spalling. Breeching occurs when the local stresses, from a close-in explosion, are so high that the full concrete thickness is punched through. Spalling occurs when tensile stresses are higher than the tensile strength of the concrete, creating fragments from the concrete cover. Breeching resistance can be increased with the use of stirrups or lacing. If breeching from a close-in explosion is to be avoided, the scaled distance ($R/W^{1/3}$) of the explosive from the structure must be at least 1.0 $ft/lb^{1/3}$, when single leg stirrups are used, or about 0.25 $ft/lb^{1/3}$ when lacing reinforcement is provided (See the Tri-Service Manual for detailed requirements). The thickness of the reinforced concrete section should also be at least equal to t_b (in).

$$t_b = 4.12(R/W^{1/3})^{-0.40}\ W^{1/3}$$

where R = distance from the center of the explosive to the structure, ft

W = explosive weight, lb. TNT equivalent

If spalling is a hazard it can be eliminated with spall plates or by using the minimum concrete thickness, t_s (in).

$$t_s = 5.31(R/W^{1/3})^{-0.40}\ W^{1/3}$$

Design Criteria for Fragment Penetration. Complete penetration of concrete (perforation) by steel fragments can be prevented by using the minimum concrete thickness, t_{pf} (in) given by the following relationship:

$$t_{pf} = 1.13\ X_f\ d^{0.1} + 1.31\ d$$

where for $x > 2d$:

$$X_f = [0.30\ W_f^{0.40}\ V_f^{1.8} + 0.575W^{0.33}](5{,}000/f'_c)^{1/2}$$

for $x \leq 2d$:

$$X_f = [0.91\ W_f^{0.37}\ V_f^{0.9}](5{,}000/f'_c)^{1/2}$$

d = diameter of fragment, in

W_f = fragment weight OZ

v_f = fragment velocity, kfps

f'_c = concrete compression strength, psi

Conclusion

The advantages of using reinforced concrete for the design of blast-hardened structures and the important recent changes to the design criteria of flexural elements have been summarized. Detailed design of hardened structures should be in accordance with the criteria in the tri-service design manual, TM 5-1300/NAVFAC P-397/AFM 88-22, "Structures to Resist the Effects of Accidental Explosions".

RECEIVED April 21, 1987

Chapter 6

Blast-Resistant Glazing

Gerald E. Meyers

Naval Civil Engineering Laboratory, Port Hueneme, CA 93043

Guidelines are presented for the design and evaluation of fixed or non-openable tempered glass windows to survive safely a prescribed blast environment described by a triangular-shaped pressure-time curve. These guidelines are in the form of load criteria for the design of both the glass panes and framing system for the window. The criteria account for both bending and membrane stresses and their effect on maximum principal stresses and the nonlinear flexural behavior of glass panes.

Historical records of explosion effects demonstrate that blast-propelled glass fragments from failed windows are often a major cause of injuries from explosions. Also, failed window glazing often leads to additional injuries as blast pressure can enter interior building spaces and subject personnel to high pressure jetting, incident overpressure, secondary debris impact and thrown body impact. These risks are heightened in modern facilities, which often have large areas of glazing.

This paper presents guidelines for the design, and evaluation, of fixed or non-openable windows to survive safely a prescribed blast environment described by a triangular-shaped pressure-time curve. Window designs using monolithic (unlaminated) thermally tempered glass based on these guidelines can be expected to provide a probability of failure equivalent to that provided by current safety standards for safely resisting wind loads.

The guidelines are presented in the form of load criteria for the design of both the glass panes and framing system for the window. The criteria account for both bending and membrane stresses and their effect on maximum principal stresses and the nonlinear behavior of glass panes. Further research is underway to extend this design criteria to both laminated tempered glass and polycarbonate.

This chapter not subject to U.S. copyright
Published 1987 American Chemical Society

Background

The design criteria for blast resistant glazing covers monolithic thermally tempered glass meeting the requirements of Federal Specifications DD-G-1403B and DD-G-451d. Additionally, thermally tempered glass is required to meet the minimum fragment weight requirements of ANSI Z97.1-1984.

Annealed glass is the most common form of glass available today. Depending upon manufacturing techniques, it is also known as plate, float or sheet glass. During manufacture, it is cooled slowly. The process results in very little, if any, residual compressive surface stress. Consequently, annealed glass is of relatively low strength when compared to tempered glass. Furthermore, it has large variations in strength and fractures into dagger-shaped, razor-sharp fragments. For these reasons, annealed glass is not recommended for use in blast-resistant windows.

Thermally tempered glass is the most readily available tempered glass on the market. It is manufactured from annealed glass by heating to a high uniform temperature and then applying controlled, rapid cooling. As the internal temperature profile relaxes towards uniformity, internal stresses are created. The outer layers, which cool and contract first, are set in compression, while internal layers are set in tension. As it is rare for flaws, which act as stress magnifiers, to exist in the interior of tempered glass sheets, the internal tensile stress is of relatively minimal consequence. As failure originates from tensile stresses exciting surface flaws in the glass, precompression permits a larger load to be carried before the net tensile strength of the tempered glass pane is exceeded. Thermally tempered glass is typically four to five times stronger than annealed glass.

The fracture characteristics of tempered glass are superior to those of annealed glass. Due to the high strain energy stored by the prestress, tempered glass will eventually fracture into small cube-shaped fragments instead of the razor-sharp, dagger-shaped fragments associated annealed glass. Breakage patterns of side and rear windows in American automobiles are a good example of the failure mode of thermally or heat-treated tempered glass.

Semi-tempered glass is often marketed as safety or heat-treated glass. However, it exhibits neither the dicing characteristic upon breakage nor the higher tensile strength associated with fully tempered glass. Semi-tempered glass is not recommended for blast-resistant windows.

Another common glazing material is wire-reinforced glass, annealed glass with an embedded layer of wire mesh. Its only use is as a fire-resistant barrier. Wire glass has the fracture and low strength characteristics of annealed glass and, although the wire binds fragments, it contributes metal fragments as an additional hazard. Wire glass is never recommended for blast-resistant windows.

Design Criteria for Glazing

Specified Glazing. The design of blast-resistant windows is currently restricted to heat-treated, fully-tempered glass in fixed or non-openable frames meeting both Federal Specification DD-G-1403B and ANSI Z97.1-1984. To preclude the possibility that stress concentrations at tong marks will cause premature failures, the glass must be tempered horizontally or in a basket. No nicks or imperfections about the edges should be permitted. Although thermally tempered glass exhibits the safest failure mode of any glass, failure under blast loading still presents a significant health hazard. Results from blast tests reveal that upon fracture, tempered glass fragments may be propelled in cohesive clumps that only fragment upon impact into smaller rock-salt-shaped fragments. Even if the tempered glass breaks up initially into small fragments, sufficient blast pressure can propel the fragments at a high enough velocity to constitute a severe danger. Because of the high likelihood of multiple edge and corner impacts by fragments of tempered glass, biomedical experts warn that the 58-ft-lb criterion for acceptable fragments should not be applied to glass. Because of these fragment dangers, blast-resistant glazing should be designed to survive with high probability its design threat.

Design Charts. Charts are presented in Figures 1 through 12 for both the design and evaluation of glazing to survive safely a prescribed blast loading with a probability of failure no greater than 0.001. The charts relate the peak blast overpressure capacity, B, of thermally tempered glazing to all combinations of the following design parameters: length/width ratio = 1.00, 1.50, 2.00, and 4.00; $1.00 \leq$ glass area ≤ 25 ft^2; $12 \leq$ width ≤ 60 inches; $2 \leq$ blast duration $\leq 1{,}000$ msec; and thickness = 1/4, 5/16, 3/8, 1/2, 5/8, and 3/4 inch (nominal). Thermally tempered glass up to 3/4 inch thick can be easily purchased in the United States. Thicknesses greater than 3/4 inch can only be obtained by lamination. Research and blast load testing are required to develop design curves with confidence for laminated glass.

Each chart has a series of curves. Each curve corresponds to the pane dimension shown to the right of the curve. Adjacent to the pane dimension is the value of B (peak blast overpressure capacity) corresponding to T = 1,000 msec. The posted value of B is intended to reduce errors when interpolating between curves.

Required Design Criteria for Frame

Sealants, Gaskets, and Beads. All gaskets or beads are required to be at least 3/8 inch wide with a Shore "A" durometer hardness of 50 and conform to ASTM Specification C509-84 (Cellular Elastomeric Preformed Gasket and Sealing Material).

The bead and sealant are required to form a weatherproof seal.

Glazing Setting. Minimum frame edge clearances, face clearance, and bite (illustrated in Figure 13) are specified in Table I.

Frame Loads. The window frame must develop the static design strength of the glass pane, r_{u}, given in Table II. Otherwise, the design is inconsistent with frame assumptions, and the peak blast pressure capacity of the window assemblies will produce a failure rate in excess of the prescribed failure rate. This results

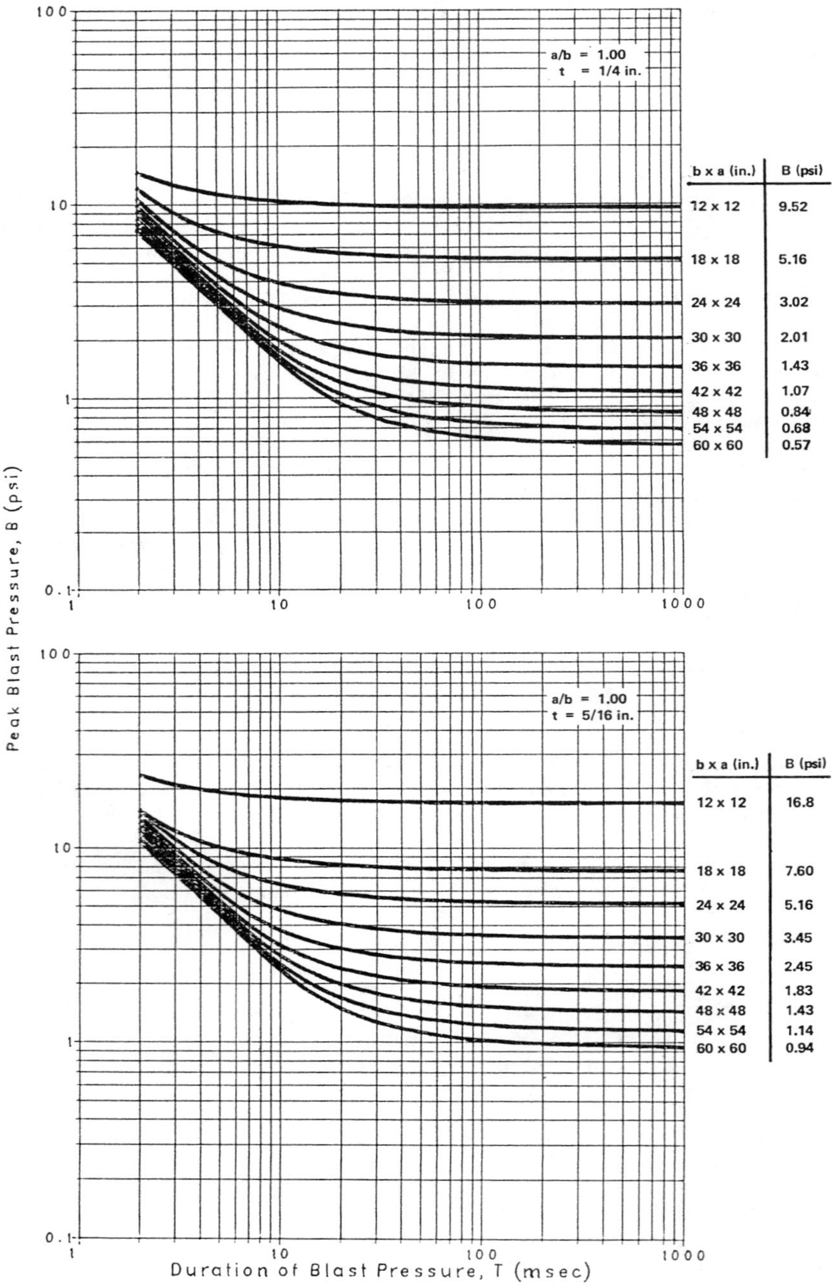

Figure 1. Peak blast pressure capacity for tempered glass panes: a/b = 1.00, t = 1/4 and 5/16 in.

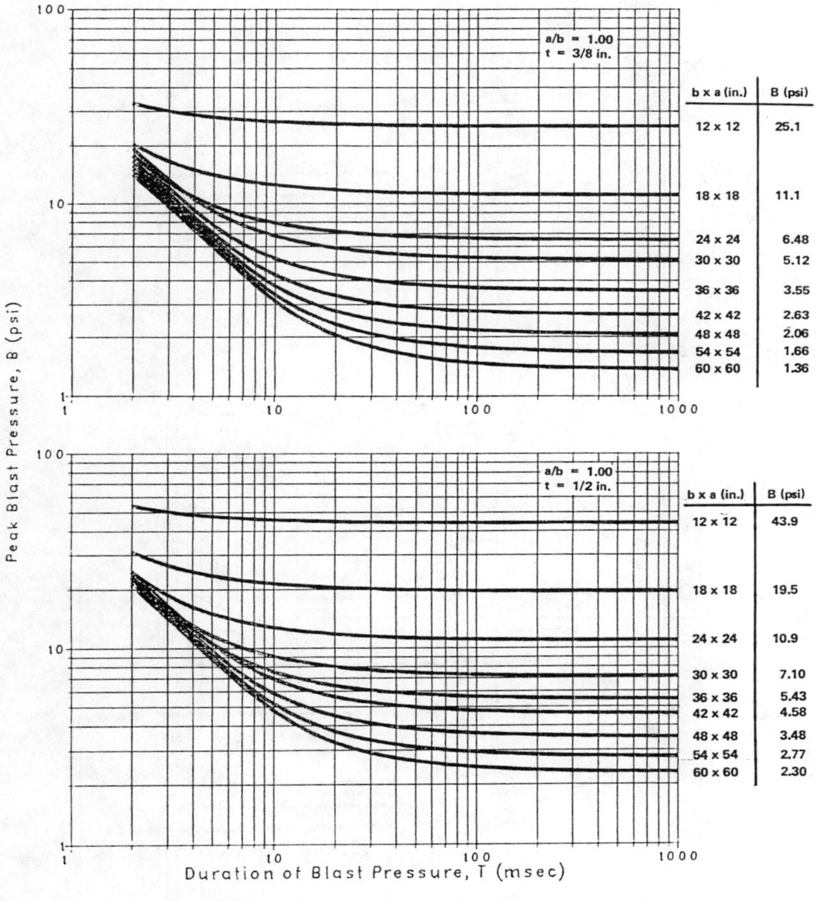

Figure 2. Peak blast pressure capacity for tempered glass panes: a/b = 1.00, t = 3/8 and 1/2 in.

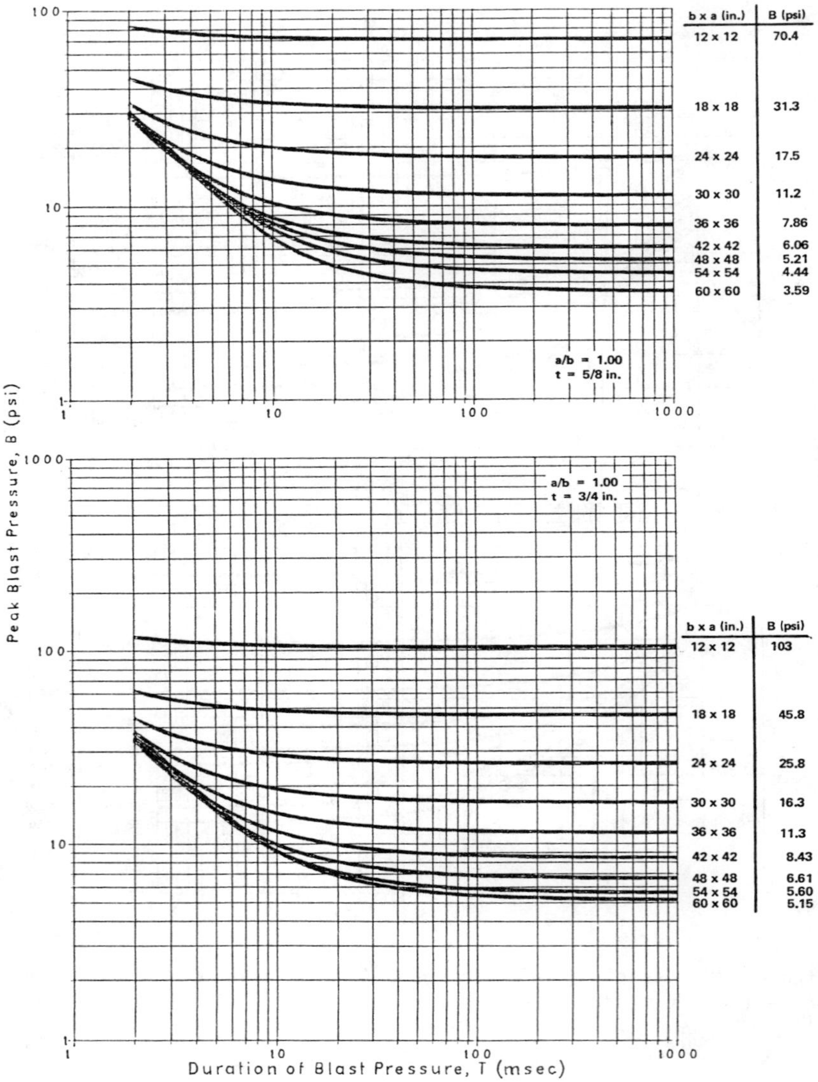

Figure 3. Peak blast pressure capacity for tempered glass panes: a/b = 1.00, t = 5/8 and 3/4 in.

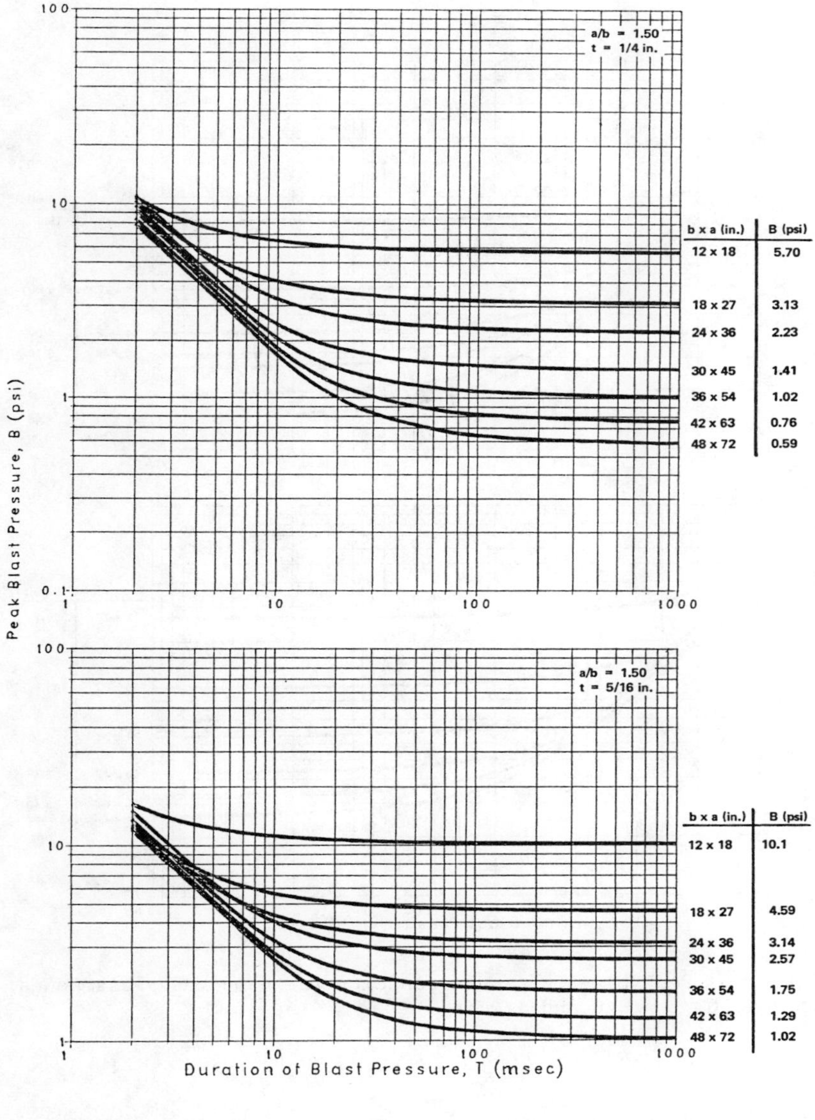

Figure 4. Peak blast pressure capacity for tempered glass panes: a/b = 1.50, t = 1/4 and 5/16 in.

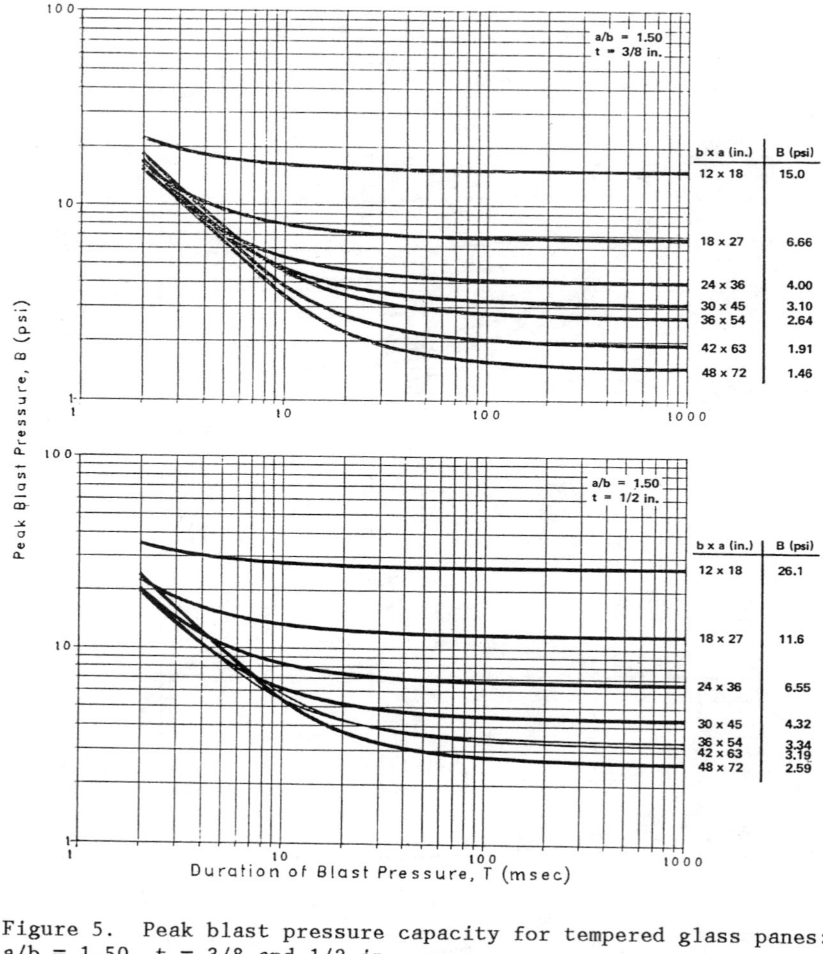

Figure 5. Peak blast pressure capacity for tempered glass panes: a/b = 1.50, t = 3/8 and 1/2 in.

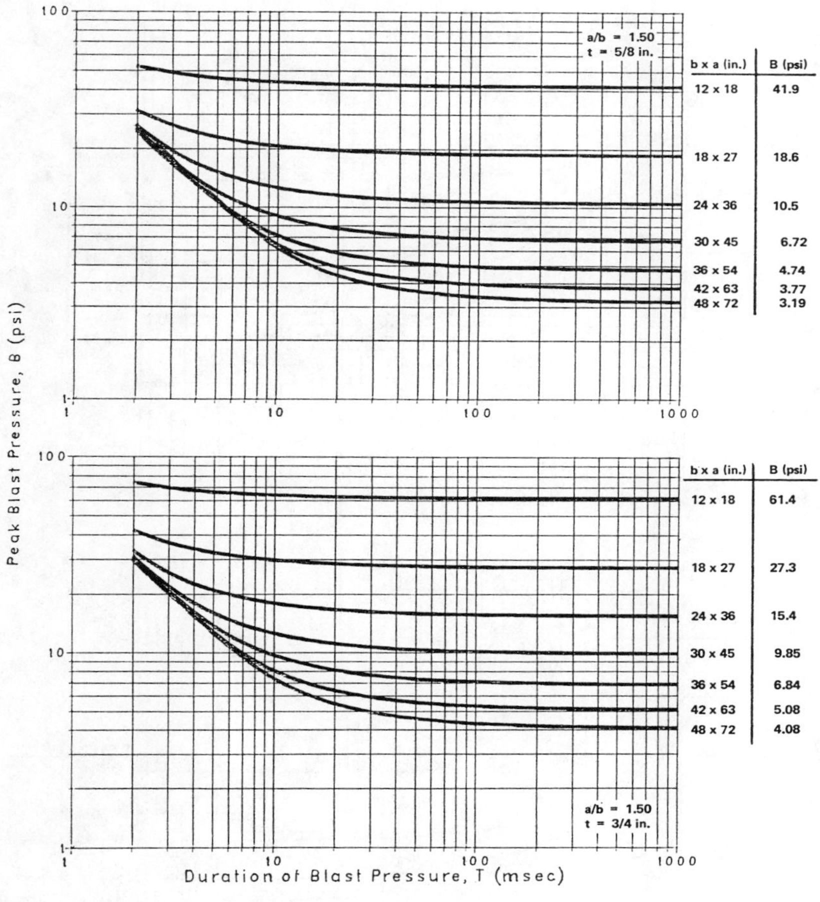

Figure 6. Peak blast pressure capacity for tempered glass panes: a/b = 1.50, t = 5/8 and 3/4 in.

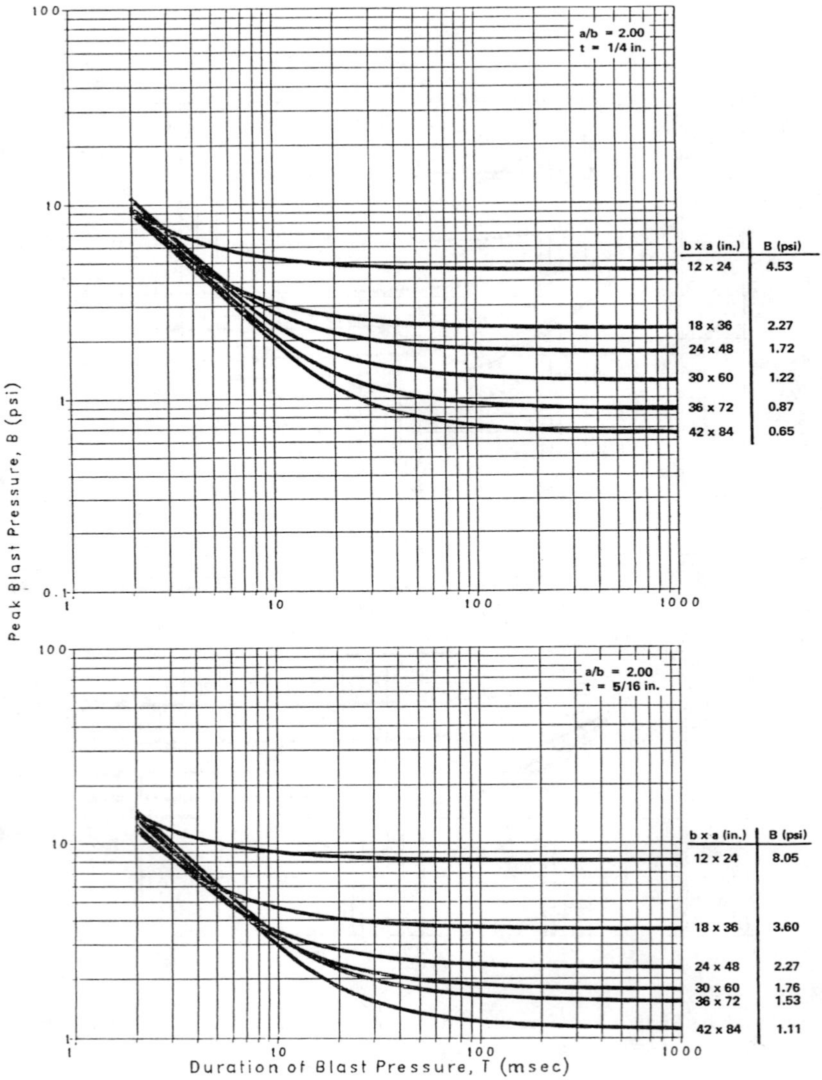

Figure 7. Peak blast pressure capacity for tempered glass panes: a/b = 2.00, t = 1/4 and 5/16 in.

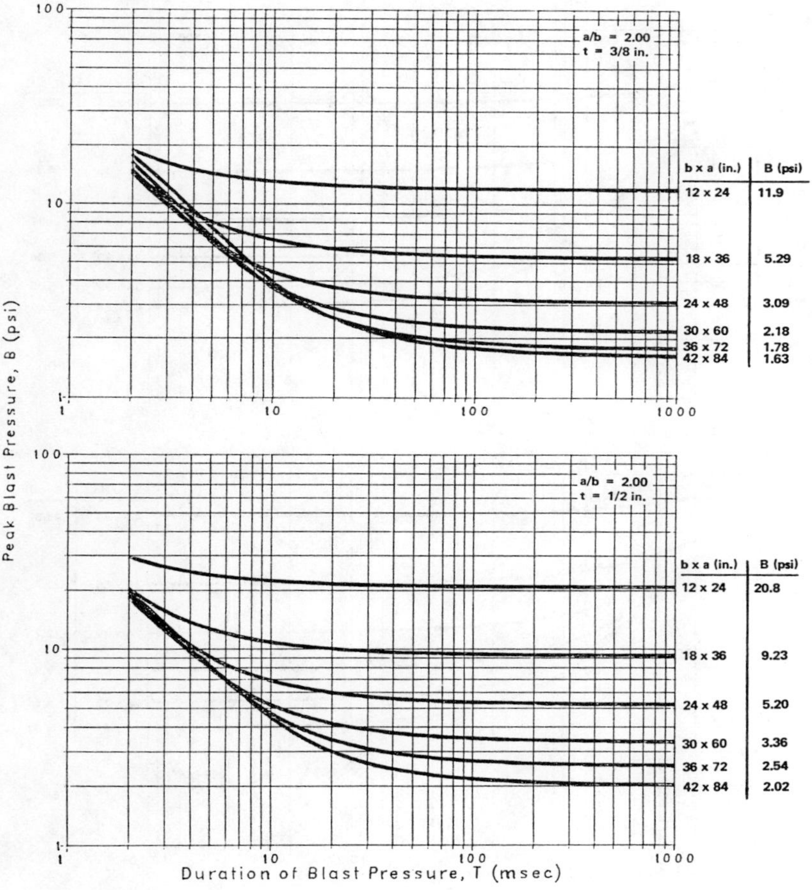

Figure 8. Peak blast pressure capacity for tempered glass panes: a/b = 2.00, t = 3/8 and 1/2 in.

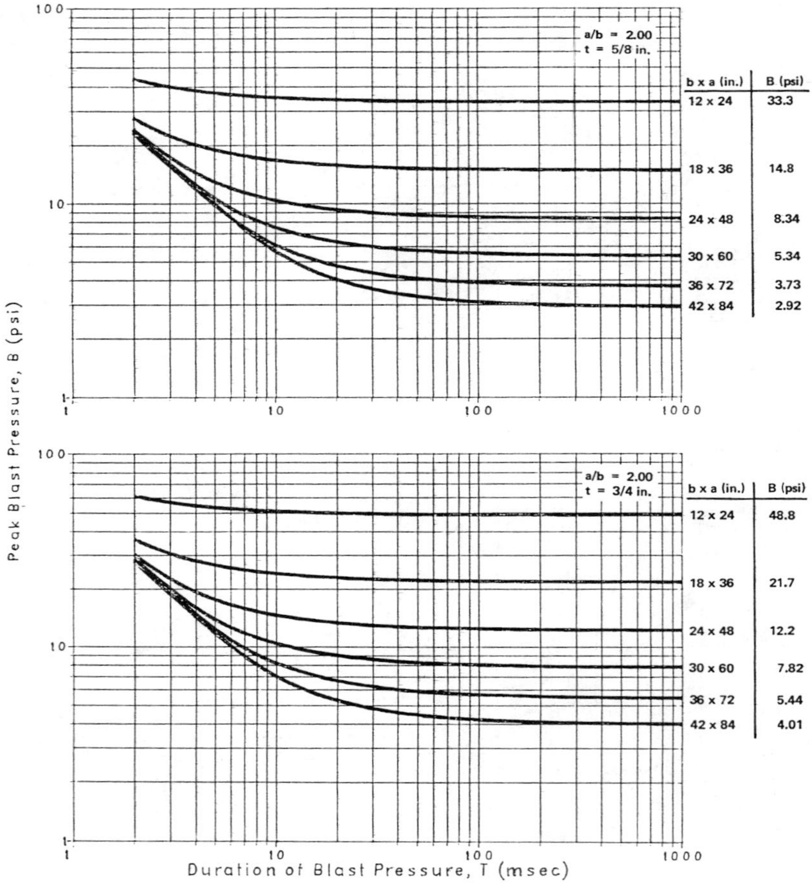

Figure 9. Peak blast pressure capacity for tempered glass panes: a/b = 2.00, t = 5/8 and 3/4 in.

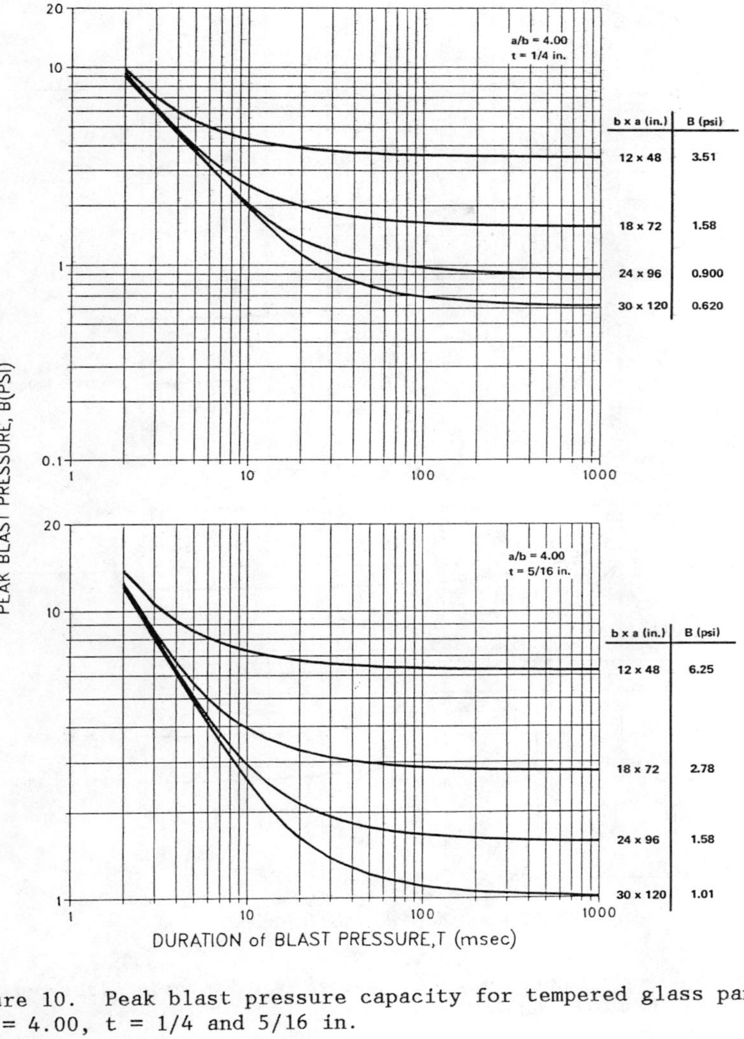

Figure 10. Peak blast pressure capacity for tempered glass panes: a/b = 4.00, t = 1/4 and 5/16 in.

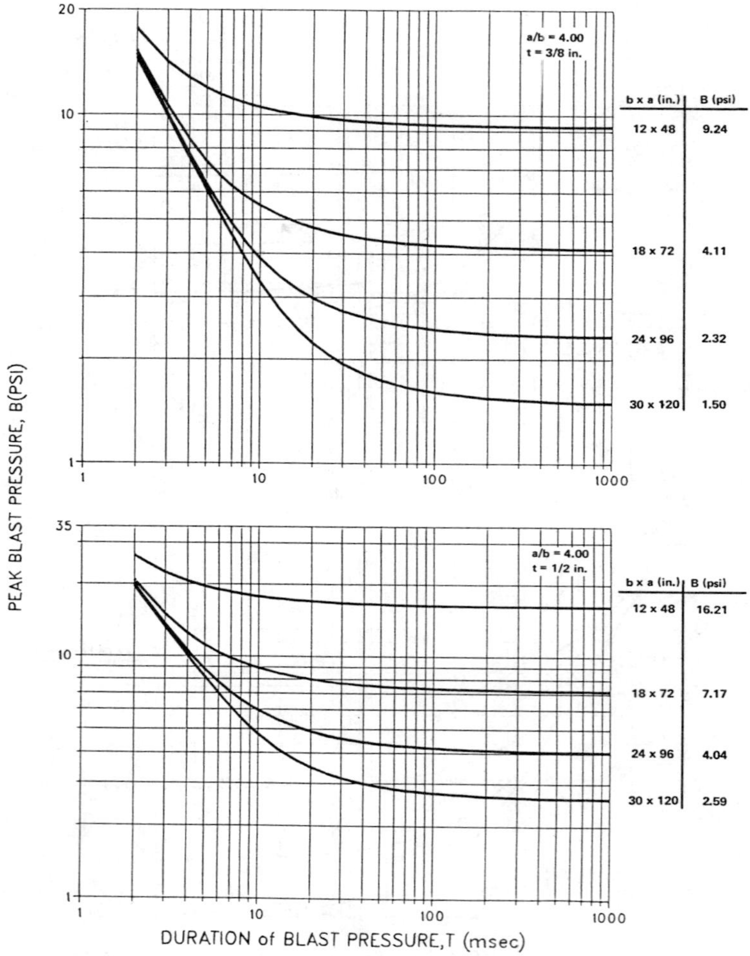

Figure 11. Peak blast pressure capacity for tempered glass panes: a/b = 4.00, t = 3/8 and 1/2 in.

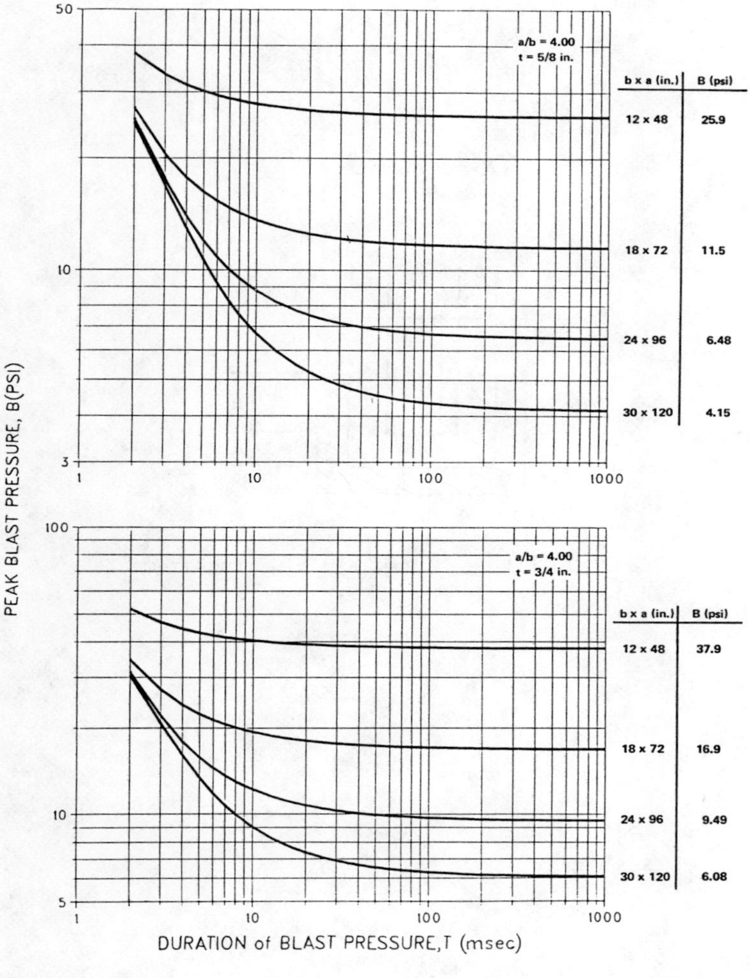

Figure 12. Peak blast pressure capacity for tempered glass panes: a/b = 4.00, t = 5/8 and 3/4 in.

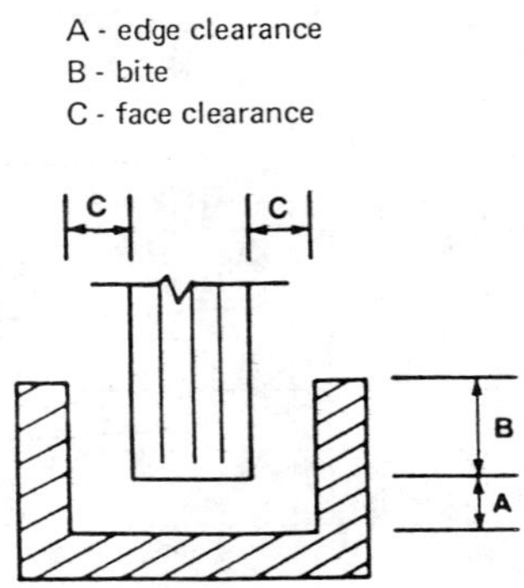

Figure 13. Edge, face, and bite requirements.

because frame deflections induce higher principal tensile stresses in the pane, thus reducing the capacity available to safely resist the blast loading.

Table I. Minimum Design Thicknesses, Clearances and Bite Requirements

Glass Thickness (Nominal) in	Glass Thickness (Nominal) mm	Actual Glass Thickness for Design, t (in)	"A" Minimum Edge Clearance (in)	"B" Nominal Bite (in)	"C" Minimum Face Clearance (in)
5/32	4.0	0.149	3/16	1/2	1/8
3/16	5.0	0.180	3/16	1/2	1/8
1/4	6.0	0.219	1/4	1/2	1/8
3/8	10.0	0.355	5/16	1/2	3/16
1/2	12.0	0.469	3/8	1/2	1/4
5/8	16.0	0.594	3/8	1/2	1/4
3/4	19.0	0.719	3/8	1/2	5/16

In addition to the load transferred to the frame by the glass, frame members must also resist the static design load, r_u, applied to all exposed members. Maximum allowable limits for frame design are:

1. <u>Deflection</u>: No frame member should have a relative displacement exceeding 1/264th of its span or 1/8 inch, whichever is less.

2. <u>Stress</u>: The maximum stress in any member should not exceed $f_y/1.65$, where f_y = yield stress of the members material.

3. <u>Fasteners</u>: The maximum stress in any fastener should not exceed $f_y/2.00$.

The design loads for the glazing are based on large deflection plate theory, but the resulting transferred design loads for the frame are based on an approximate solution of small deflection theory for normally loaded plates. Analysis indicates this approach to be considerably simpler and more conservative than using the frame loading based exclusively on large deflection plate behavior, characteristic of window panes. The effect of the static design load, r_u, applied directly to the exposed frame members of width, w, is also considered. The design load, r_u, produces a line shear, V_x, applied by the long side, a, of the pane equal to:

$$V_x = C_x r_u b \sin(\pi x/a) + r_u w, \text{ lb/in} \qquad (1)$$

The design load, r_u, produces a line shear, V_y, applied by the short side, b, of the pane equal to:

Table II. Static Design Strength, r_u(psi), for Tempered Glass*

[a = long dimension of glass pane (in.); b = short dimension of glass pane (in.)]

ASPECT RATIO = 1.00

Glass Size, b x a (in.)	Static Design Strength (psi) for a Window thickness, t, of --					
	3/4 in.	5/8 in.	1/2 in.	3/8 in.	5/16 in.	1/4 in.
12x12	206	141	87.7	50.3	27.5	20.2
13x13	176	120	74.7	42.8	23.9	17.6
14x14	151	103	64.5	36.9	21.1	15.5
15x15	132	90.1	56.1	32.2	18.7	14.2
16x16	116	79.2	49.3	28.3	16.7	13.4
17x17	103	70.1	43.7	25.5	15.1	12.7
18x18	91.6	62.5	39.0	23.1	14.1	12.6
19x19	82.2	56.1	35.0	21.0	13.5	12.1
20x20	74.2	50.7	31.6	19.2	12.9	11.0
21x21	67.3	46.0	28.6	17.7	12.7	10.0
22x22	61.3	41.9	26.4	16.3	12.6	9.20
23x23	56.1	38.3	24.4	15.1	11.8	8.52
24x24	51.5	35.2	22.7	14.3	10.9	7.91
25x25	47.5	32.4	21.2	13.8	10.1	7.43
26x26	43.9	30.0	19.7	13.4	9.39	7.00
27x27	40.7	27.9	18.5	12.9	8.80	6.62
28x28	37.9	26.2	17.4	12.8	8.26	6.22
29x29	35.3	24.6	16.4	12.6	7.78	5.86
30x30	33.0	23.2	15.4	12.6	7.39	5.53
31x31	30.9	21.9	14.6	12.0	7.04	5.22
32x32	29.0	20.8	14.2	11.3	6.71	4.94
33x33	27.4	19.7	13.8	10.6	6.39	4.69
34x34	26.0	18.7	13.5	10.0	6.07	4.45
35x35	24.8	17.8	13.2	9.50	5.77	4.23
36x36	23.6	17.0	12.8	9.05	5.50	4.04
37x37	22.5	16.2	12.7	8.63	5.24	3.86
38x38	21.5	15.4	12.7	8.24	5.01	3.69
39x39	20.5	14.8	12.6	7.88	4.79	3.53
40x40	19.7	14.4	12.5	7.57	4.58	3.39
41x41	18.8	14.1	11.9	7.30	4.39	3.25
42x42	18.1	13.8	11.4	7.04	4.21	3.12
43x43	17.3	13.5	10.9	6.80	4.05	3.00
44x44	16.7	13.2	10.4	6.56	3.90	2.89
45x45	16.0	13.0	9.99	6.32	3.75	2.78
46x46	15.4	12.9	9.59	6.08	3.62	2.68
47x47	14.9	12.8	9.24	5.86	3.49	2.58
48x48	14.5	12.7	8.91	5.65	3.37	2.49
49x49	14.2	12.6	8.59	5.45	3.25	2.41
50x50	14.0	12.6	8.30	5.27	3.15	2.33
51x51	13.7	12.4	8.02	5.09	3.04	2.25
52x52	13.5	11.9	7.76	4.92	2.95	2.18
53x52	13.3	11.5	7.54	4.76	2.85	2.11
54x54	13.1	11.1	7.33	4.61	2.77	2.05
55x55	12.9	10.7	7.13	4.47	2.68	1.99
56x56	12.8	10.3	6.94	4.33	2.60	1.93
57x57	12.7	9.99	6.76	4.20	2.53	1.87
58x58	12.7	9.66	6.59	4.08	2.45	1.82
59x59	12.6	9.38	6.40	3.97	2.38	1.77
60x60	12.6	9.11	6.22	3.85	2.32	1.72

*Panes to the right and below the stepped dividing line behave according to large deflection plate theory.

Table II. Continued.

[a = long dimension of glass pane (in.); b = short dimension of glass pane (in.)]

ASPECT RATIO = 1.50

Glass Size, b x a (in.)	Static Design Strength (psi) for a Window Thickness, t, of --					
	3/4 in.	5/8 in.	1/2 in.	3/8 in.	5/16 in.	1/4 in.
12x18	123	83.8	52.3	29.9	16.3	11.9
13x19.5	105	71.4	44.5	25.5	13.9	10.5
14x21	90.2	61.6	38.4	22.0	12.3	9.43
15x22.5	78.6	53.6	33.4	19.2	11.1	8.88
16x24	69.1	47.2	29.4	16.8	10.0	8.26
17x25.5	61.2	41.8	26.0	14.9	9.31	8.14
18x27	54.6	37.3	23.2	13.3	8.85	8.02
19x28.5	49.0	33.4	20.8	12.3	8.84	7.90
20x30	44.2	30.2	18.8	11.4	7.83	7.78
21x31.5	40.1	27.4	17.1	10.6	7.81	7.62
22x33	36.5	24.9	15.6	9.86	7.80	7.03
23x34.5	33.4	22.8	14.2	9.32	7.77	6.45
24x36	30.7	21.0	13.1	8.98	7.77	5.95
25x37.5	28.3	19.3	12.4	8.64	7.63	5.50
26x39	26.2	17.9	11.7	8.24	7.19	5.10
27x40.5	24.3	16.6	11.0	7.86	6.69	4.74
28x42	22.6	15.4	10.4	7.85	6.24	4.42
29x43.5	21.0	14.4	9.89	7.85	5.83	4.14
30x45	19.7	13.4	9.42	7.84	5.47	3.88
31x46.5	18.4	12.8	9.16	7.83	5.13	3.64
32x48	17.3	12.2	8.91	7.82	4.83	3.43
33x49.5	16.2	11.6	8.65	7.72	4.55	3.27
34x51	15.3	11.1	8.34	7.62	4.30	3.13
35x52.5	14.4	10.6	8.05	7.28	4.07	3.00
36x54	13.6	10.2	8.02	6.90	3.85	2.87
37x55.5	13.0	9.78	7.99	6.55	3.66	2.74
38x57	12.5	9.42	7.96	6.22	3.47	2.61
39x58.5	12.0	9.21	7.93	5.92	3.33	2.50
40x60	11.6	9.01	7.91	5.64	3.21	2.39
41x61.5	11.2	8.82	7.88	5.38	3.09	2.29
42x63	10.8	8.60	7.85	5.13	2.98	2.19
43x64.5	10.4	8.35	7.77	4.91	2.88	2.10
44x66	10.1	8.12	7.69	4.70	2.77	2.02
45x67.5	9.71	7.90	7.62	4.50	2.66	1.94
46x69	9.42	7.69	7.35	4.31	2.56	1.86
47x70.5	9.25	7.62	7.06	4.14	2.47	1.79
48x72	9.08	7.55	6.78	3.97	2.38	1.73

*Panes to the right and below the stepped dividing line behave according to large deflection plate theory.

Continued on next page.

Table II. Continued.

[a = long dimension of glass pane (in.); b = short dimension of glass pane (in.)]

ASPECT RATIO = 2.00

Glass Size, b x a (in.)	Static Design Strength (psi) for a Window Thickness, t, of --					
	3/4 in.	5/8 in.	1/2 in.	3/8 in.	5/16 in.	1/4 in.
12x24	97.6	66.6	41.5	23.8	13.0	9.05
13x26	83.1	56.7	35.4	20.3	11.0	7.81
14x28	71.7	48.9	30.5	17.5	9.52	6.87
15x30	62.4	42.6	26.6	15.2	8.31	6.29
16x32	54.9	37.5	23.4	13.3	7.43	5.81
17x34	48.6	33.2	20.7	11.9	6.72	5.40
18x36	43.4	29.6	18.5	10.6	6.26	5.03
19x38	38.9	26.6	16.6	9.49	5.86	4.71
20x40	35.1	24.0	14.9	8.56	5.51	4.56
21x42	31.9	21.7	13.6	7.85	5.19	4.46
22x44	29.0	19.8	12.4	7.25	4.90	4.42
23x46	26.6	18.1	11.3	6.73	4.64	4.39
24x48	24.4	16.6	10.4	6.39	4.55	4.37
25x50	22.5	15.3	9.56	6.08	4.47	4.32
26x52	20.8	14.2	8.84	5.79	4.40	4.24
27x54	19.3	13.2	8.23	5.53	4.39	4.01
28x56	17.9	12.2	7.73	5.29	4.38	3.74
29x58	16.7	11.4	7.27	5.07	4.37	3.50
30x60	15.6	10.7	6.86	4.86	4.31	3.28
31x62	14.6	9.98	6.57	4.67	4.25	3.09
32x64	13.7	9.36	6.32	4.58	4.08	2.93
33x66	12.9	8.80	6.08	4.52	3.85	2.78
34x68	12.2	8.31	5.87	4.47	3.64	2.64
35x70	11.5	7.91	5.66	4.41	3.44	2.51
36x72	10.8	7.53	5.47	4.40	3.26	2.39
37x74	10.3	7.18	5.29	4.39	3.11	2.28
38x76	9.73	6.86	5.12	4.38	2.97	2.18
39x78	9.24	6.62	4.96	4.37	2.84	2.08
40x80	8.78	6.42	4.81	4.34	2.72	1.98
41x82	8.37	6.23	4.67	4.30	2.60	1.89
42x84	8.03	6.05	4.60	4.25	2.50	1.80

ASPECT RATIO = 4.00

Glass Size, b x a (in.)	Static Design Strength (psi) for a Window Thickness, t, of --					
	3/4 in.	5/8 in.	1/2 in.	3/8 in.	5/16 in.	1/4 in.
12x48	75.7	51.7	32.2	18.5	10.1	7.02
13x52	64.5	44.0	27.5	15.7	8.57	5.99
14x56	55.6	38.0	23.7	13.6	7.39	5.16
15x60	48.5	33.1	20.6	11.8	6.43	4.52
16x64	42.6	29.1	18.1	10.4	5.66	3.99
17x68	37.7	25.8	16.1	9.20	5.01	3.56
18x72	33.7	23.0	14.3	8.20	4.49	3.19
19x76	30.2	20.6	12.9	7.36	4.05	2.87
20x80	27.3	18.6	11.6	6.65	3.67	2.60
21x84	24.7	16.9	10.5	6.03	3.34	2.37
22x88	22.5	15.4	9.59	5.49	3.06	2.18
23x92	20.6	14.1	8.77	5.03	2.81	2.02
24x96	18.9	12.9	8.05	4.63	2.59	1.88
25x100	17.5	11.9	7.42	4.28	2.39	1.76
26x104	16.1	11.0	6.86	3.97	2.23	1.66
27x108	15.0	10.2	6.36	3.70	2.09	1.57
28x112	13.9	9.49	5.92	3.45	1.96	1.49
29x116	13.0	8.85	5.52	3.22	1.84	1.41
30x120	12.1	8.27	5.15	3.02	1.75	1.34

*Panes to the right and below the stepped dividing line behave according to large deflection plate theory.

$$V_y = C_y r_u b \sin(\pi y/b) + r_u w, \text{ lb/in} \quad (2)$$

The design load, r_u, also produces a corner concentrated load, R, tending to uplift the corners of the window pane equal to:

$$R = C_R r_u b^2, \text{ lb} \quad (3)$$

Distribution of these forces as loads acting on the window frame is shown in Figure 14. Table III presents the design coefficients, C_x, C_y, and C_R for practical aspect ratios of the window pane. Linear interpolation can be used for aspect ratios not presented.

Table III. Coefficients for Frame Loading

a/b	C_R	C_x	C_y
1.00	0.065	0.495	0.495
1.10	0.070	0.516	0.516
1.20	0.074	0.535	0.533
1.30	0.079	0.554	0.551
1.40	0.083	0.570	0.562
1.50	0.085	0.581	0.574
1.60	0.086	0.590	0.583
1.70	0.088	0.600	0.591
1.80	0.090	0.609	0.600
1.90	0.091	0.616	0.607
2.00	0.092	0.623	0.614
3.00	0.093	0.644	0.655
4.00	0.094	0.687	0.685

Although frames with mullions are included in the design criteria, it is recommended that single pane frames be used.

Experience indicates that mullions complicate the design and reduce reliable fabrication of blast-resistant frames. If mullions are used, the loads given by Equations 1, 2, and 3 should be used to check the frame mullions and fasteners for compliance with the deflection and stress criteria stated above.

Special design consideration should be taken so that the deflection of the building wall will not impose deflections on the frame greater than 1/264th of the length of the edge of the pane. Where it is impossible to achieve enough building wall rigidity, it is recommended that the frames be pinned at the corners to the structure in a manner to isolate the frame from wall rotation.

Rebound. Response to the dynamic blast load, will cause the window to rebound with a negative (outward) deflection. The outward pane displacement and the stresses produced by the negative deflection must be safely resisted by the window while positive pressures act on the window. Otherwise, the window which safely resists stresses

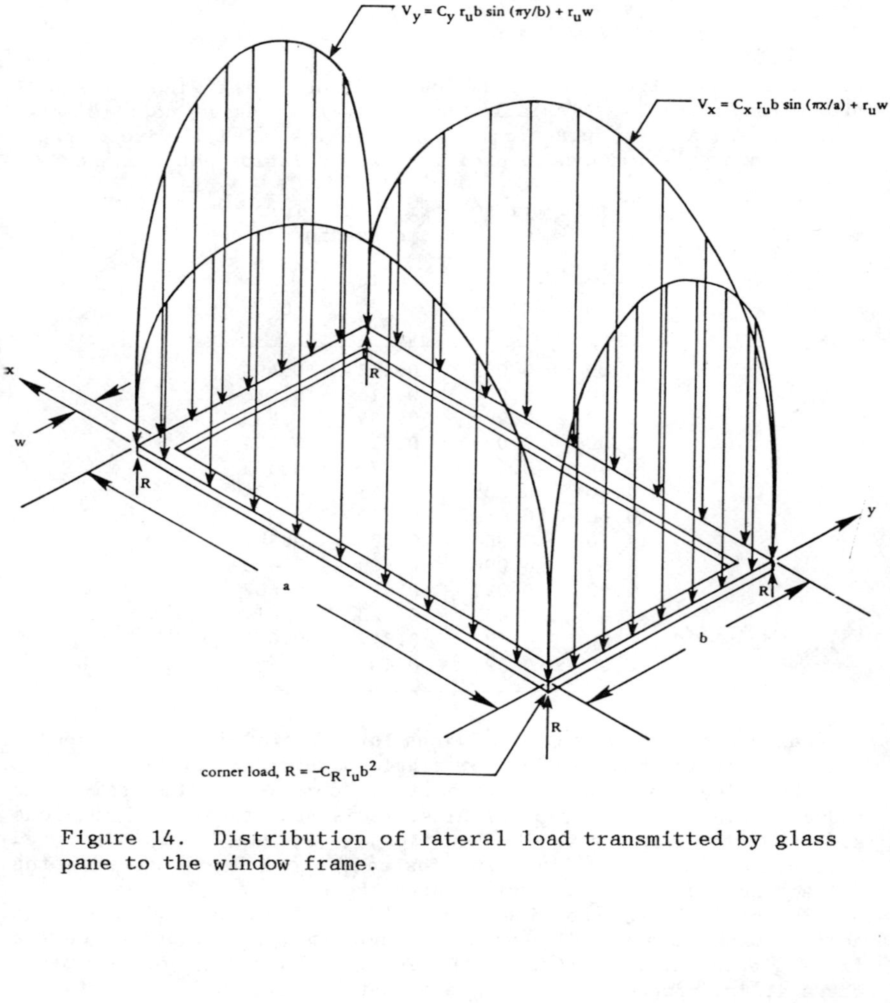

Figure 14. Distribution of lateral load transmitted by glass pane to the window frame.

induced by positive (inward) displacements may fail in rebound while the positive pressure still acts. This can propel glass fragments into the interior of the structure. However, if the window fails in rebound during the negative (suction) phase of the blast loading, glass fragments will be drawn away from the structure. If glass failure does not present a hazard to personnel outside the structure, glass may be permitted to fail during the negative load phase. Rebound will occur during the negative load phase if the effective blast duration, T, is no greater than one half the natural period of vibration, T_n, of the glass pane. For $T \geq 10\ T_n$, significant rebound does not occur during the positive blast pressure phase. Therefore, rebound can be neglected as a design consideration. For $0.5 < T/T_n < 10$, the frame must be designed for the peak negative resistance occurring during the positive overpressure phase.

Installation Inspection

A survey of glazing failures due to wind load indicates that improper installation of setting blocks, gaskets or lateral shims, or poor edge bite is a significant cause of failure because of the resultant unconservative support conditions. To prevent premature glass failure, a strenuous quality control program is required.

Literature Cited

Meyers, G.E. Design Criteria and Test Acceptance Specification for Blast Resistant Windows, Department of Defense Explosive Safety Seminar, Anaheim, CA, Aug 1986.

RECEIVED May 13, 1987

Chapter 7

Interim Design Criteria for Polycarbonate Blast-Resistant Glazing

Gerald E. Meyers and James E. Tancreto

Naval Civil Engineering Laboratory, Port Hueneme, CA 93043

Glazing is often the weakest element in the protective capability of a structure against blast, fragments, and ballistics. Polycarbonate and glass-clad polycarbonate can overcome these deficiencies. This paper establishes credible and reliable interim design values for blast resistant glazing utilizing polycarbonate as a structured layer. Required design of the frame and edge engagement or bite of the glazing are also included.

Glazing is often the weakest element in the protective capability of a structure against blast, fragments and ballistics. Over the last few years, the U.S. Naval Civil Engineering Laboratory has developed and validated design charts and tables for thermally tempered glass (Reference 1 and 2) for use where blast overpressure is the predominate threat. However, this glass does not provide a comparable level of protection against fragments, ballistics, or forced entry. Also, even if a laminated thermally tempered glass remains intact after fragment or ballistic impact, it will lose both its transparency and operational effectiveness.

Polycarbonate and glass clad polycarbonate can overcome these deficiencies. As a glazing material, it has established a long track record against fragments, ballistics and physical assault. However, no design method or practice existed to guide the reliable design of polycarbonate to resist blast. It is the intent of this paper to fill this immediate and pressing need and to establish credible and reliable interim design values for blast resistant glazing utilizing polycarbonate as a structured layer. Required design of the frame and edge engagement or bite of the glazing are also included as they are requisite for a successful blast resistant design.

While conservative engineering assumptions have been employed, a large data base yet needs to be developed to validate the presented design. However, the limited testing in the engineering literature, even at high overpressures, provides initial confidence in the present designs. Also, the dynamic or blast analysis used to generate

This chapter not subject to U.S. copyright
Published 1987 American Chemical Society

the design charts are independent from those used to create design tables for the physical security setting such as in Reference 3. The close correspondence between the solution methodology employed for this paper (numerical integration of the differential equations of motion) and that used for the physical security design tables (the response spectra solution of an equivalent linear elastic spring-mass model) are mutually confirmatory.

Material Characteristics

Polycarbonate is a thermoplastic and is often marketed under tradenames such as Lexan or Tuffak. It should not be confused with acrylic plastics, marketed under tradenames such as Plexiglas or Lucite, which are flammable and exhibit a brittle failure mode.

Polycarbonate is available monolithically (in a single sheet) in thicknesses up to 1/2 inch. In this range of thickness, polycarbonate is twice as expensive as thermally tempered glass. In thicknesses over 1/2 inch where lamination is required, it is roughly three times as expensive as an equivalent thermally tempered laminated lite.

Other than cost, polycarbonate's main disadvantage is that it experiences greater environmental degradation than glass, especially due to the effects of ultraviolet radiation and abrasion. However, chemical coatings, such as Lexan's MARGARD or Tuffak's CM3, are available to provide some protection from abrasion. Ultraviolet inhibitors are also available for most commercial polycarbonate. Greater protection against both abrasion and ultraviolet attack is afforded by encapsulating the polycarbonate in glass. Incidentally, this will enhance both the ballistic and chemical resistance of the glass. Unfortunately, testing of older glass-clad polycarbonate indicates that even glass-encapsulated polycarbonate with ultraviolet inhibitors will suffer degradation of load carrying and penetration resistance over time. In recognition of this fact and to be conservative, this paper will assume a reduced maximum stress for polycarbonate and not employ the potential benefits of ductile or post-elastic yield design.

Pane Design Theory

A maximum flexural stress of 9,500 psi is assumed for polycarbonate. This conservative stress value should account for degradation in ultraviolet stabilized polycarbonate exposed to long term solar exposure. While more research is required in this area, it is reasonable to expect at least a ten year useful life for ultraviolet stabilized polycarbonate. A Young's modulus of 345,000 psi and a Poisson's ratio of 0.38 are also assumed for polycarbonate.

The polycarbonate glazing is modeled as a simply supported plate subjected to nonlinear center deflections up to 15 times the pane thickness. Using the finite element solution of Moore (Reference 4), the resistance function is generated for each pane under consideration. Typically, the resistance is concave up, as illustrated for typical pane sizes in Figure 1. This occurs because membrane stresses induced by the stretching of the neutral axis of the pane become more pronounced as the ratio of the center pane deflection to the pane

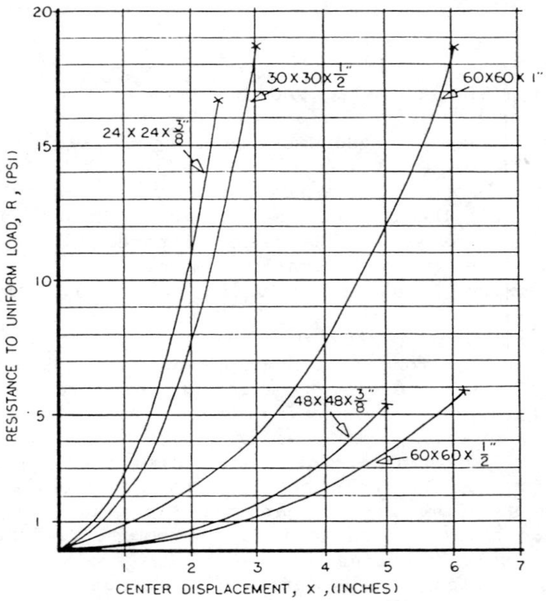

Figure 1. Resistance function of polycarbonate.

thickness increases. In a few cases of thin panes with long spans where the center deflection associated with a maximum stress of 9,500 psi in the plate exceeds 15 pane thicknesses, a smaller design maximum stress associated with a 15-pane thickness is chosen. This limitation both restricts the solution to the valid range of the Von Karmen equations used by the finite element program to develop the resistance function and the practical edge engagement developed by commercially available frames.

A single-degree-of-freedom approach is used to perform the dynamic or blast analysis. The resistance function is modeled as five linear segments and a Wilson-Theta numerical integration of the equation of motion is performed. A maximum time step of integration smaller than 1/25th of the natural period of vibration of the corresponding segment of the resistance function is used. No damping is assumed and the effective mass of the pane is limited by a load mass factor between 0.63 and 0.79 depending upon the aspect ratio (ratio of pane length to width).

The blast load is modeled as a triangular-shaped overpressure time curve. The blast overpressure rises instantaneously to the peak overpressure, B, then decays linearly with a blast pressure duration, T. The pressure is uniformly distributed over the surface of the plate and is applied perpendicular to the pane.

Monolithic action is assumed between adjoining polycarbonate layers for the following reasons. First, recent static load testing at the Naval Civil Engineering Laboratory indicates this to be a good assumption. Second, the large deflections experienced by the relatively flexible polycarbonate means that a relatively high proportion of load is being carried in membrane action rather than bending. Interlaminar shear capacity between plates does not affect this very efficient mode of structural capacity. Finally, it is anticipated that the high strain rates associated with blast loading will further increase the shear capacity of most, if not all, interlaminar plastics in current commercial use.

To prevent failure due to the disengagement of the pane out of the frame, bite or edge engagement depths are required. They are based upon the assumption that the plate will distort as a spheroid surface. At the maximum design center deflection of 15 pane thicknesses, this conservatively approximates the deflection shape function. To be conservative, a 0.5-inch safety margin is added to all calculations.

Glazing Design Charts

Figures 2 through 9 are design charts for ultraviolet stabilized polycarbonate under blast load. Charts are provided for pane thicknesses of 1/4, 3/8, 1/2, and 1 inch for pane areas up to 25 ft at pane aspect ratios (pane length to width ratios) of 1.00, 1.50, 2.00 and 4.00. The charts relate the peak experienced blast overpressure capacity, B, for convenient pane dimensions across the spectrum of encountered blast durations. Depending on the orientation of the window to the charge, the blast overpressure may either be incident or reflected. The pane dimensions (measured across the span from the gasket centerline) peak blast capacity at 1000 msec, B, static frame design pressure, r_u, and the required bite are printed to the right

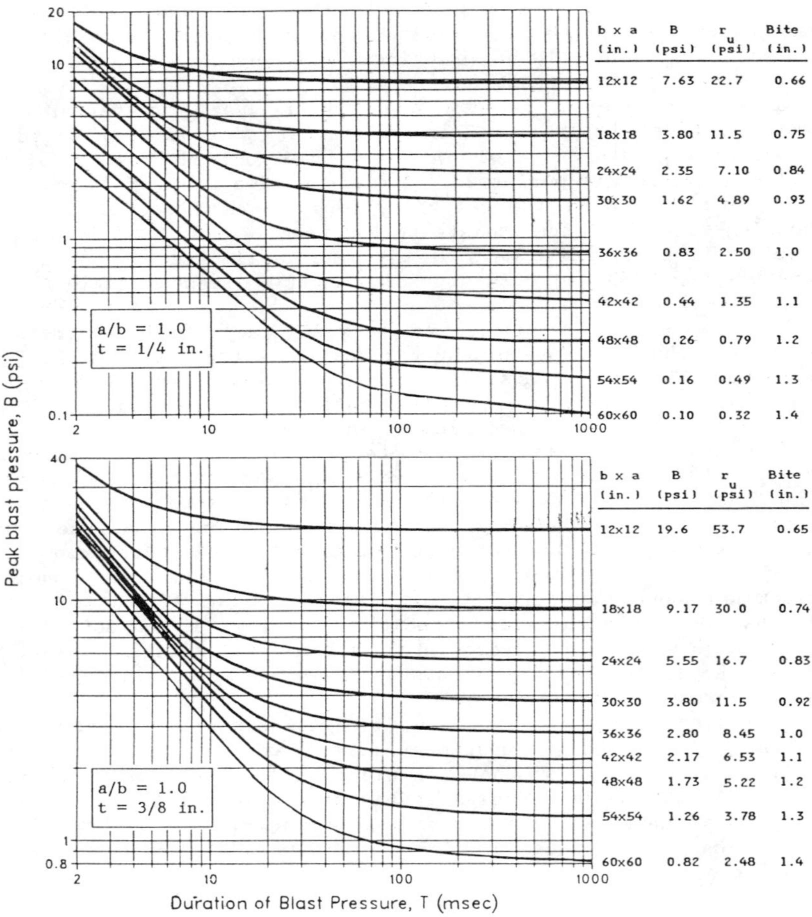

Figure 2. Peak blast pressure capacity for polycarbonate: a/b = 1.0; t = 1/4 and 3/8 in.

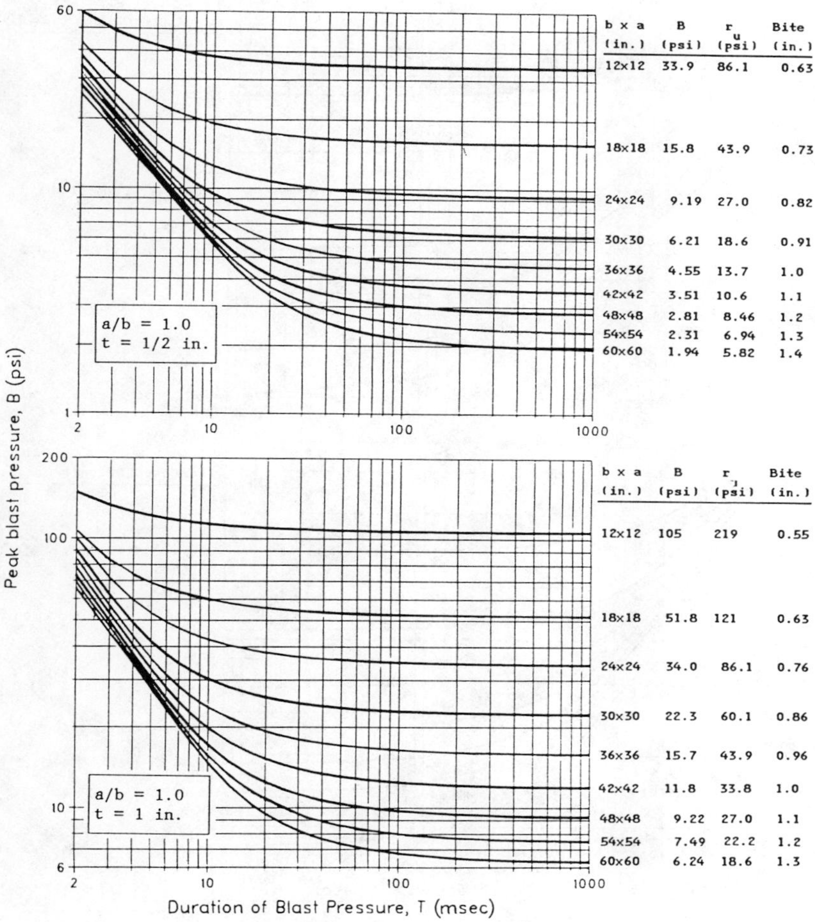

Figure 3. Peak blast pressure capacity for polycarbonate: a/b = 1.0; t = 1/2 and 1 in.

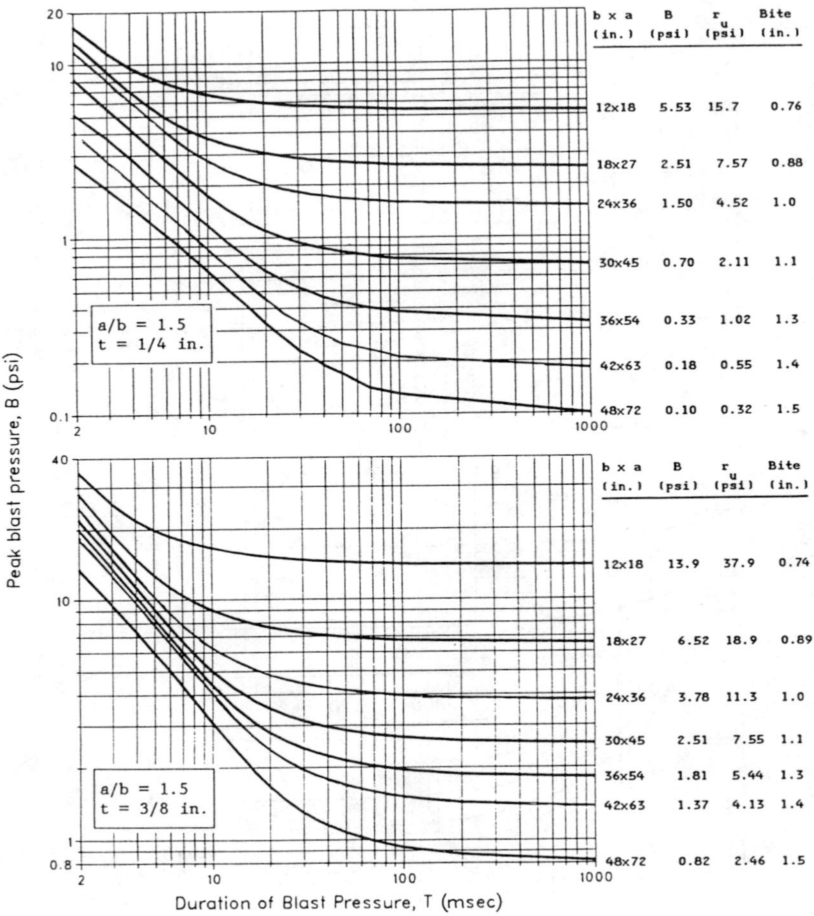

Figure 4. Peak blast pressure capacity for polycarbonate: a/b = 1.5; t = 1/4 and 3/8 in.

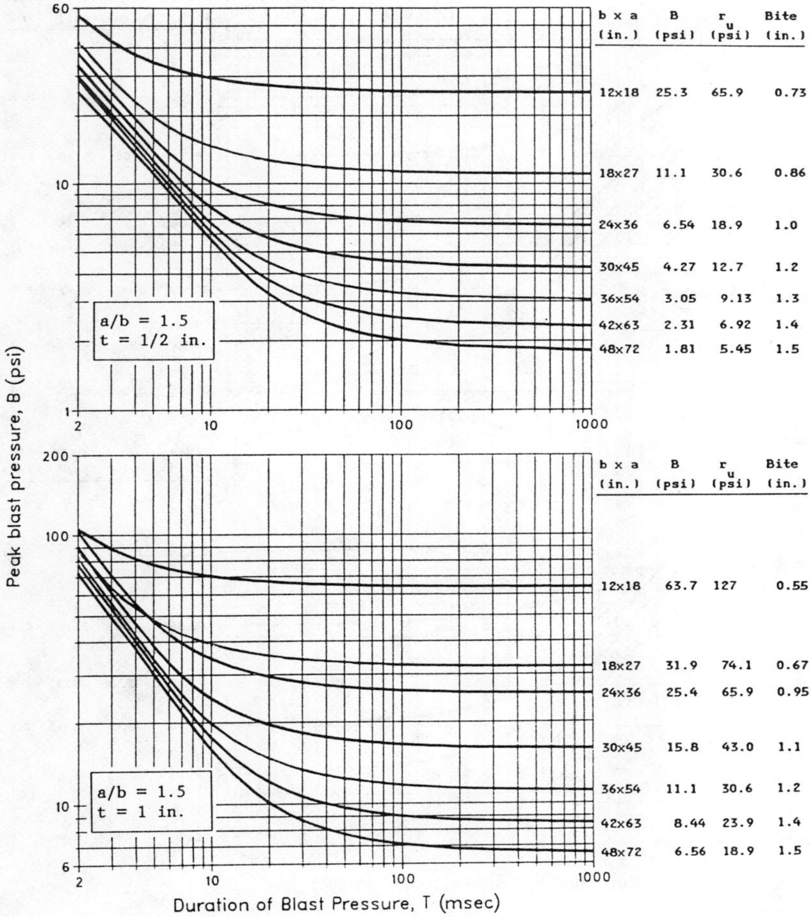

Figure 5. Peak blast pressure capacity for polycarbonate: a/b = 1.5; t = 1/2 and 1 in.

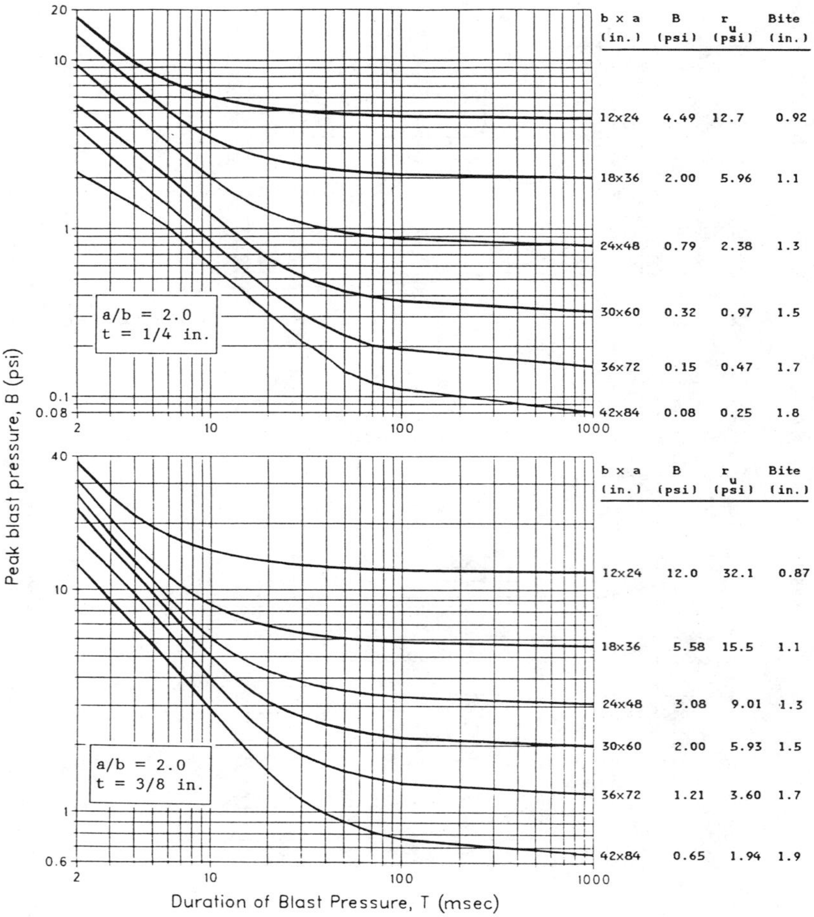

Figure 6. Peak blast pressure capacity for polycarbonate: a/b = 2.0; t = 1/4 and 3/8 in.

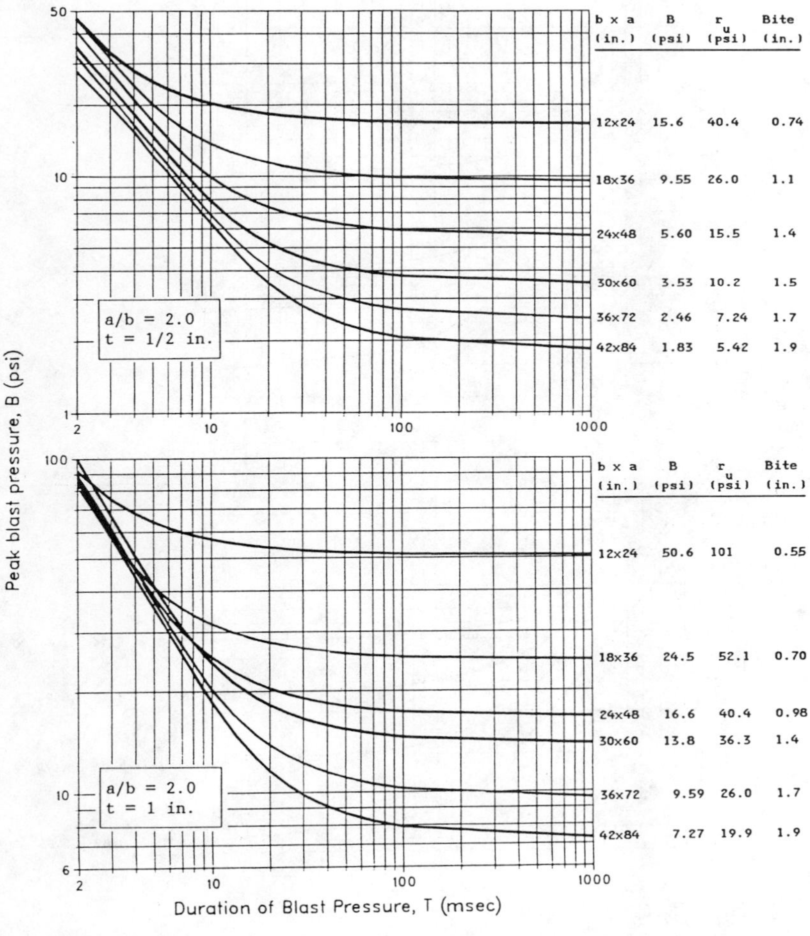

Figure 7. Peak blast pressure capacity for polycarbonate: a/b = 2.0; t = 1/2 and 1 in.

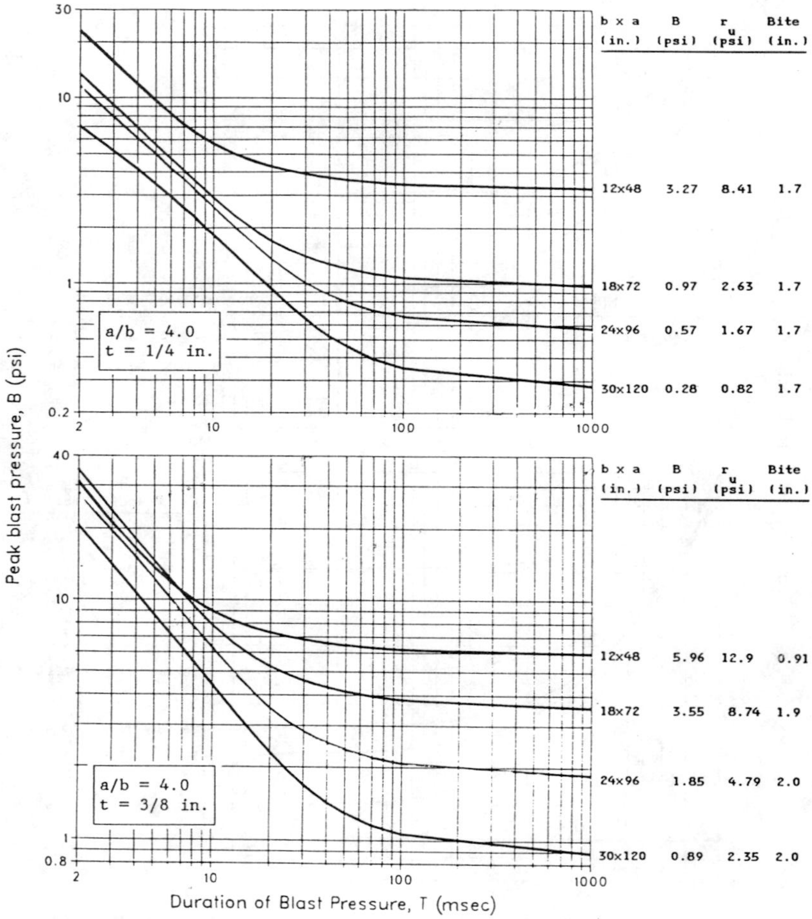

Figure 8. Peak blast pressure capacity for polycarbonate: a/b = 4.0; t = 1/4 and 3/8 in.

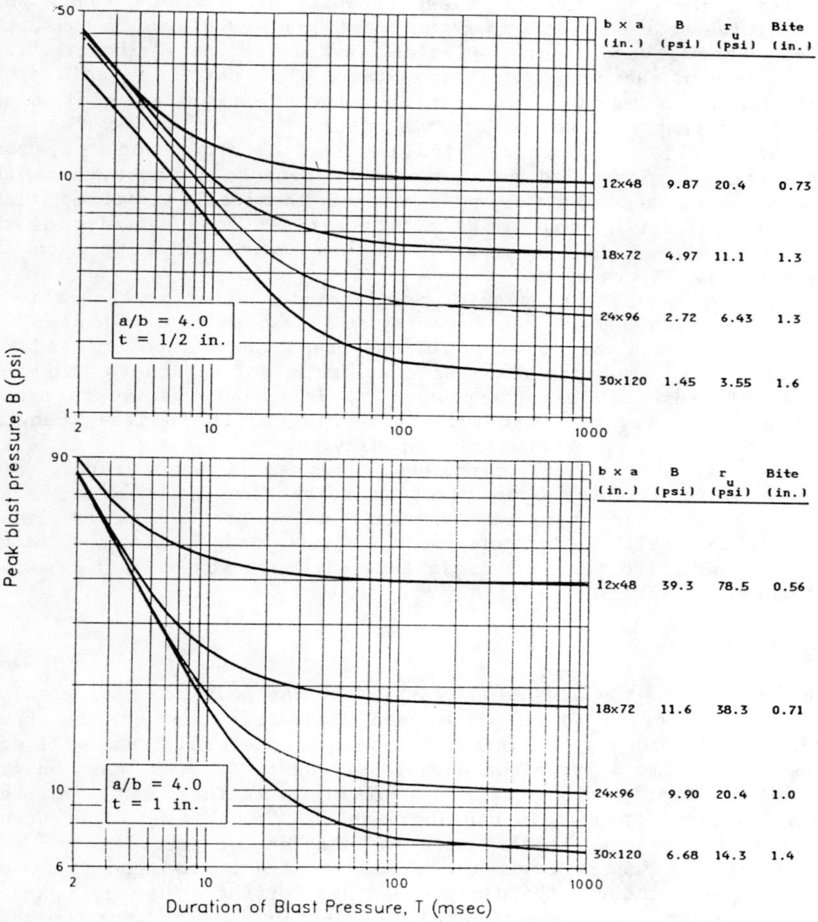

Figure 9. Peak blast pressure capacity for polycarbonate: a/b = 4.0; t = 1/2 and 1 in.

of each design curve. To reflect current manufacturing tolerances and to be conservative, design thickness used to calculate blast capacities were limited to 95% of the nominal thickness.

It is worth noting that blast capacity of a polycarbonate pane is sensitive to the duration of the blast load. Because of this, the typical short overpressure duration testing of polycarbonate with small close-in charges with frame set-ups that permit a rapid pressure clearing time may give an unconservative estimate of blast capacity in many real world threat scenarios.

Engineering judgment is also required in assessing the blast capacity of a glass-clad polycarbonate. Because in most cases the annealed, semi-tempered, or sodium-based chemically tempered glass does not contribute substantially to the blast load capacity of the cross section, it is conservative to base blast capacity upon the polycarbonate layers alone.

In many cases, the dynamic amplification factor or the ratio of static load to dynamic load capacity will exceed two. This is because of the concave up shape of the resistance function and the mobilization of membrane resistance at large deflection to thickness ratios. Because of this phenomenon, it is unconservative to assume the blast capacity of polycarbonate glazing to be no less than one half of its static pressure load capacity.

At very short blast durations, some small area 1-inch thick panes exhibit slightly less blast capacity than panes with larger areas. This occurs because the small panes are acting as linear plates with small deflections under blast loads while the larger panes can mobilize membrane resistance without exceeding the maximum design stress of 9,500 psi.

Frame Requirements

To be effective, the blast load carried by the polycarbonate glazings must be transferred to the frame and ultimately through the structure. If not properly designed, the pane or pane and frame will disengage and become a large and dangerous fragment. Also, care must be taken to properly design the supporting structure for the frame loads. Failure to do this can increase the probability of structure collapse. This is especially true in retrofit construction.

While the design loads for the panes are based upon large deflection plate theory, the design loadings for the frame are based on an approximate solution of small deflection theory for normally loaded plates. Analysis indicates this approach to be considerably simpler and more conservative than using the frame loading based exclusively on large deflection plate behavior. The effect of the static design load, r_u, applied directly to the exposed frame members of width, w, should also be considered. The design load, r_u, produces a line shear, V_x, applied by the long side, a, of the pane equal to:

$$V_x = C_x r_u b \sin(\pi x/a) + r_u w, \qquad \text{lb/in.} \qquad (1)$$

The design load, r_u, produces a line shear, V_y, applied by the short side, b, of the pane equal to:

$$V_y = C_y r_u b \sin(\pi y/b) + r_u w, \qquad \text{lb/in.} \qquad (2)$$

The design load, r_u, also produces a corner concentrated load, R, tending to uplift the corners of the window pane equal to:

$$R = C_R r_u b^2, \qquad \text{lb} \qquad (3)$$

Distribution of these forces as loads acting on the window frame is shown in Figure 10. Static frame design loads, r_u, are provided for each pane in the third column of the design data to the right of each design chart. Table I presents the design coefficients, C_x, C_y, and C_R for practical aspect ratios of the pane. Linear interpolation can be used for aspect ratios not presented. Frame deflections should be limited to no more than 1/100 the length of the supporting span. This is a significant benefit compared to the more rigid restrictions associated with tempered glass.

Although frames with mullions are covered in the design criteria, it is recommended that single pane frames be used. Experience indicates that mullions complicate the design and reduce reliable fabrication of blast-resistant frames.

Table I.

a/b	C_R	C_x	C_y
1.00	0.065	0.495	0.495
1.50	0.085	0.581	0.574
2.00	0.092	0.623	0.614
4.00	0.094	0.687	0.685

Frame Bite

Minimum frame bites or frame edge engagements are required for polycarbonate to provide enough edge support to carry the blast load and prevent pane disengagement. The fourth column to the right of each design chart presents the required bite for each pane.

Rebound

Response to the dynamic blast load will cause the window to rebound with a negative (outward) deflection. The outward pane displacement and the stresses produced by the negative deflection must be safely resisted by both the pane and frame. If operational requirements dictate an operational window after the blast, the frame, connections, and wall should be designed to also resist the static frame design load, r_u, in the outward direction. If the window can be permitted to fail after the positive blast pressure has decayed, more economical frames can be used, as the negative static design load can be reduced to 0.67 of r_u. For blast durations greater than 250 msec, significant rebound does not occur during the positive pressure phase.

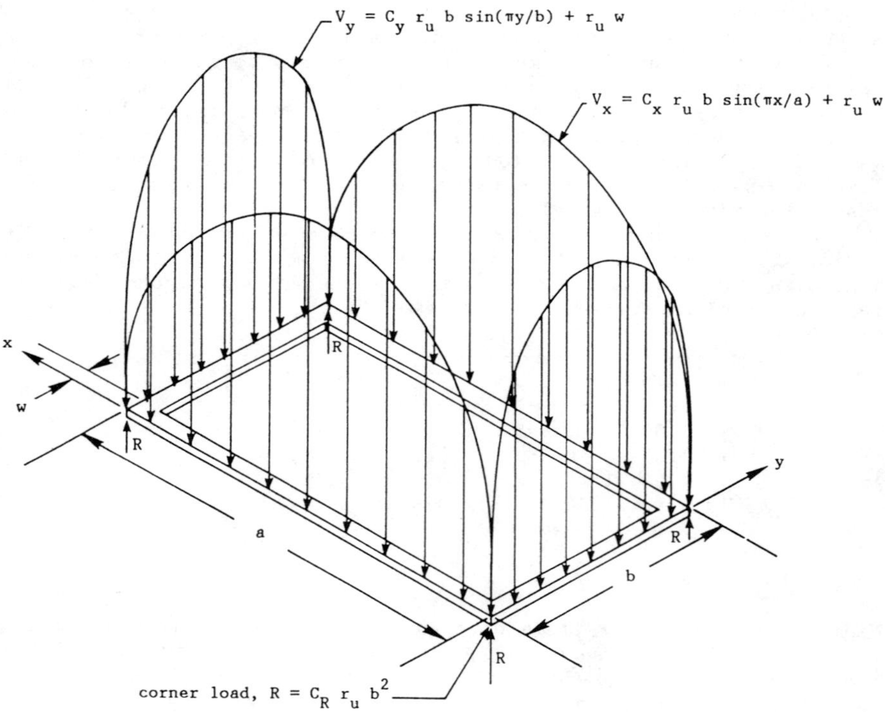

Figure 10. Frame design loading to be applied by the pane to the frame.

Literature Cited

1. Meyers, G.E. "Design Criteria and Acceptance Test Specification for Blast Resistant Thermally Tempered Glass Glazing," Department of Defense Explosive Safety Seminar, Anaheim, CA, Aug 1986.

2. Physical Security: Shore Station Engineering Design Criteria, (MIL-HDBK-1013/1), (to be published).

3. Physical Security - Planning and Requirement Guide. U.S. Army Corps of Engineers, Omaha, Neb, Mar 1987.

4. Moore, J.M. FSA Task Report No. 5101-291: Thickness Sizing of Glass Plates Subjected to Pressure Loads. Jet Propulsion Laboratory, Pasadena, CA, Aug 1982.

RECEIVED April 21, 1987

THERMAL EFFECTS

Chapter 8

Thermal Effects: An Overview

W. R. Herrera and L. M. Vargas

Southwest Research Institute, 6220 Culebra Road, Post Office Drawer 28510, San Antonio, TX 78284

The design of a facility to withstand the thermal effects of an accidental ignition and subsequent combustion must address a number of safety related design requirements and considerations including thermal safety requirements. It is an accepted fact that 80 to 90 percent of the combustible contents of building compartments are consumed during the period of a fully developing fire. Fire safety design considerations must therefore be selected to lessen the danger of spread of fire, smoke and toxic materials beyond the confines of the fire compartment. Special design features as well as detection and suppression devices can be selected to provide control and even extinguishment of potential fires. Unfortunately, a fire can reach a fully developed stage (according to experts, one in twenty incidents) thus requiring that the design of all facilities perform satisfactorily during a full fire scenario.

Once a fire in a compartment has reached the fully developed stage, chances of saving personnel trapped in the compartment or equipment within the compartment are very low. The principal design effort must therefore be directed toward providing life safety and minimization of property loss in communicable areas within the facility. In order to accomplish these goals, the designer must identify the potential damage mechanisms and their effects as well as estimating the magnitude of these threats. The designer can then develop the necessary corrective actions to protect personnel and equipment be it through safe siting of facilities or through personnel protection schemes. A very helpful tool to a designer for identifying both the potential hazards and controls is the hazards/risk assessment analysis.

This chapter of the book presents papers dealing with the identification and mitigation of potential hazards to personnel and facilities, with the development of personnel protection schemes and with the safe siting of facilities.

The following brief paragraphs have been included to provide the reader with a more detailed explanation of the types of safety related analysis techniques that a designer must address.

0097-6156/87/0345-0148$06.00/0
© 1987 American Chemical Society

Risk Assessment for Operations

A designer, as part of his facility design analysis, should perform a hazards analysis or risk assessment of the various processes which will be conducted within the facility in order to determine what potential thermal dangers or threats exist to personnel and equipment. A hazards analysis or risk assessment will provide for the identification of potential hazards and of the necessary corrective actions/measures to prevent or control the hazard. Early in the design of a facility, the processes and equipment may be conceptual and at this stage, a preliminary hazards analysis can be performed. It is early in the design that a preliminary hazards analysis can be most helpful because its implementation will have little impact on schedules and will provide the largest potential for cost savings. As the designs are modified and refined, the hazards analyses should reflect all changes in order to insure that all potential hazards and risks have been identified and that the corresponding controls have been implemented. This iterative loop should be continued throughout the design and construction phase of the facility as well as throughout the identification of processes and the installation of the process equipment. It must be emphasized that throughout the performance of the hazards analysis or risk assessment the primary emphasis is on personnel safety. Processes or equipment indentified as potential hazards should be modified, re-designed or re-evaluated in order to insure a safe system.

Expected Effects and Damage Mechanisms

Design practices stem from standard fire test procedures in which the temperature history of the test furnace is regarded as an index of the destructive potential of a fire. Thus, the practice of describing the expected effects and damage mechanism is based on temperature histories. This standard design practice is convenient but lacks accuracy in terms of structural performance. The severity of a fire should address the expected intensity of the heat flux that will impact the structure and the duration of heat penetration. A simple analysis of the expect nature of an unwanted fire can be based on the heats of combustion and pyrolysis of the principal contents in the facility. The heat of combustion will identify the destructive nature of the fire, while the heat of pyrolysis will identify the severity of the fire within the compartment itself and will also identify the destructive potential of the fire in adjacent spaces.

Prediction of Thermal Exposure Magnitude

Harmathy (1) provides a convenient way of characterizing "real world" fires in terms of three fire severity parameters:

1) The overall penetration flux, $\bar{q}$ (Watts/meter^2), i.e., heat flux absorbed by the compartment boundaries, averaged spatially over the boundary surfaces, and temporarily over the period of full development;

2) The duration of a fully developed fire, T (seconds); and,

3) The average temperature of the compartment gases, $\bar{T}g$ (average "fire" temperature), °K, averaged over the compartment volume and temporaily over the period of full development.

Energetic materials such as pyrotechnics and propellants undergo rapid exothermic decomposition or reaction in contrast to industrial materials that are classed as fire hazardous. Design considerations have to address processing conditions that identify the chemical and physical states of the ignitible material, material of fabrication with the ignitable material contacts, quantities and temperatures involved, and the likelihood that these conditions will promote transition to an explosive reaction after ignition. If one precludes the potential for an unwanted explosive reaction, the nature of the fire will be such that the penetrating flux, $\bar{q}$, will be a function of the quantity of material present in the compartment. The duration of the fire, T, will be in the range of fractions of a second to seconds with the average temperature of the compartment gases, Tg, reaching a saturation level that is dependent on the rate of heat release, $\dot{q}$, (watts/sec/sec) and the mass burning rate, $\dot{M}$ (kg/sec).

Herrera, Vargas, et al. (2) report experimental measurements of the behavior of energetic materials burning in a compartment. The results indicate that as the critical loading density, M_c (kg/m^3) increases, the mass burning rate inside the compartment reaches a steady state condition and unburned material is carried out in the plume. Burning of the unburned material then takes place outside the compartment, thereby contributing to the destructive potential of the fire in adjacent spaces.

Thermal Protection Siting Criteria

The siting of facilities housing hazardous processes or materials is another method that a designer can use to improve personnel safety. It is the intent of every designer to design a safe facility, however, a designer should consider the potential for accident occurrence and design the facility such that the quantity of materials that could potentially become involved is minimized. Limiting involvement can be accomplished by safe siting of buildings and process bays within the buildings. Siting of facilities for storage of munitions, propellants and explosives has been regulated for some time and siting criteria is well documented in reports such as AMC-R-385-100 (3). This criteria is geared more for fragment impact and blast loading of structures than for thermal loads since the primary threats that would result from an ignition in this type of a storage facility would be fragments and blast. In siting a facility for thermal loads, the designer must concern himself with a number of additional factors including: the flame spread rate, the potential fire surface area, the effect of confinement on the fire, and the firebrands that could develop and would continue to propagate the fire to adjacent buildings. Once the designer has estimated the size, duration and intensity of the potential fireball, then he can establish "safe" distances between buildings. Another technique for reducing the potential for fire communication and spread is to identify those areas within a facility where the potential for fires exists and then isolating these areas or bays. For example, if a mixing bay has the potential for a large fire, then similar mixing bays should not be located adjacent to one another to prevent propagation from one bay to the next. If the bays must be placed adjacent to one another, then precautions should be taken to isolate each bay from the other using full size dividing walls, fire proof doors, deluge systems, water curtains, etc.

Personnel Protection Requirements

The design of a facility which will handle hazardous materials requires that the designer concern himself with the potential threats to personnel working in the facility. These threats can be of a thermal nature, i.e., fires or explosions, or the threats can be of a chemical or toxic nature. Personnel can be protected from these various threats in several ways: the process operations can be mechanized, thereby eliminating any operator exposure to the hazards; the process operation itself can be desensitized, thereby making an accident less probable; fire detection and suppression systems can be installed in hazardous areas; and the operators can be equipped with the necessary protective clothing, air supplies, etc. needed to shelter or isolate the operator from a dangerous situation. Of the aforementioned protection schemes, personnel protective clothing is the easiest "fix" to implement. Recently, great strides have been made in the development of both thermal and chemical protective clothing with the clothing being not only safe but also fairly comfortable to wear.

Literature Cited

1. Harmathy, T. Z. ASTM Special Technical Publication No. 685 1979, DBR Paper No. 854, 78 p.

2. Herrera, W. R., et al. "A Study of Fire Hazards from Combustible Ammunition: Effects of Scale and Confinement (Phase II)," SwRI Final Report No. 01-7327 for DOD Explosives Safety Board, Contract MDA903-82-C-0526 (December 1984).

3. AMC Regulation, AMC-R 385-100, Safety Manual, HQ, U.S. Army Material Command, (August 1985).

RECEIVED March 23, 1987

Chapter 9

Remote Mixing and Handling Procedures for Pyrotechnic Materials

Thomas E. Shook[1], Loy M. Aikman[1], Max Frauenthal[1], David Garcia[1], Joe G. Janski[1], and F. L. McIntyre[2]

[1]Pine Bluff Arsenal, Pine Bluff, AR 71602-9500
[2]NASA National Space Technology Laboratories (NSTL), Sverdrup Technology Inc., Technical Services Laboratory, NSTL Station, MS 39529

Proper safety testing and classification of pyrotechnic energetic capacity will allow the selection of modern remote equipment for the manufacture of pyrotechnic material in a safe and economical manner. Examples are provided for new remote mixing/handling equipment certified to handle pyrotechnic mixtures at a Department of Defense facility located at Pine Bluff Arsenal (PBA), Arkansas. Safety testing is described along with a suggested rationale for improving productivity, safety, and manufacturing costs.

Pyrotechnic materials are reported to have taken many lives since the beginning of recorded history, especially where scale up from small batches has occurred. In recent times several works have appeared (1,2) which have provided an adequate description of the chemistry of pyrotechnics. The formulator, having both inadequate process equipment and lack for a rational laboratory test and classification system of energetic capacity, has been forced to handle pyrotechnic materials in small batches. This required personnel using equipment designed in the 1940's and 1950's to perform the labor intensive functions of weighing, grinding, mixing, feeding and compaction in close proximity to hazardous materials that resulted in a high degree of risk.

In the past ten years the chemical industry, primarily pharmaceuticals, has demanded more efficient and safer methods for mixing and granulating of solid systems to carry various doses of drugs in well mixed blends. The result was that the pyrotechnic industry could obtain and modify commercial Jet Air Mixers, Fluidized Bed mixers/Driers (Glatt), and MIGRAD (Mixer-Granulator-Dryer) mixers along with air transport of solids from weigh feeders to the process mixers. Modern fill and press equipment were also developed which resulted in remote systems for the entire manufacturing process that was free from close proximity to the operator and provided a significant reduction in personnel.

To take advantage of this process equipment for pyrotechnics and expand production to large volume remote systems has required the

This chapter not subject to U.S. copyright
Published 1987 American Chemical Society

development of a rational classification of energetic capacity to predict the levels of material which may be safely handled. Pyrotechnics are usually included in the definition of explosives (3). The same quantity/distances values apply for a delay mixture as they would for equal amounts of TNT (4). The energetics of a pyrotechnic are often not the same and provisions for proper in-process and final classification should be based upon the energetics of the material. Benefits gained from proper classification of pyromixtures other than possible increased productivity and safety are in the planning of pyrotechnic production facilities such that proper separation of buildings and equipment may be enjoyed without a large cost in acreage. Elimination of the construction of expensive blast resistant buildings may also be avoided. The Department of Defense (DOD) has established procedures (Table I) for in-process Safety Classification.

Table I. DOD Safety Classification Tests

PARAMETRIC:	SENSITIVITY:
Autoignition Temperature:	Card Gap Test:
Decomposition Temperature:	Detonation Test:
Explosion Temperature:	Electrical Spark Sensitivity:
Apparent Bulk Density:	Ignition and Unconfined Burning:
Fuel/Oxidizer Ratio:	Impact Sensitivity:
Gas Volume:	Friction Sensitivity (Roto-Friction Test Device)
Heat of Combustion:	
Heat of Reaction:	OUTPUT:
Propagation Index:	Burn Time: (Cube).
	Burn Time: ("Vee" block)
STABILITY:	Pressure Time:
Thermal Stability:	a) Peak Pressure
Vacuum Drying Weight loss	b) Time to Peak
	c) Rate of Rise

Additional tests that are required include mass effects and full scale "worst-case" tests. These allow for the assignment of an in-process interim qualification downgrading classification from 1.1 to 1.3 or 1.4 (Table II). Initially the pyrotechnic engineer selects the equipment of choice to meet production requirements relative to quantity of material, ease of operations, and apparent safety considerations. Highly reactive mixtures (starter mixes, flares etc.) are usually in class 1.1 or 1.2, which limits the quantity to 100 lbs. Less reactive materials like screening smokes and signals may be mixed in larger quantities. Mass effect and detonation tests are not required in class 1.1 and 1.2 since a small amount of material (100 lbs) exists in the system. These tests are required to gain classification as class 1.3 or 1.4 and are carried out on batch sizes ranging from 200 to 2,000 lbs. These tests include detonation, deflagration, shock, flame, and explosive charge. A pressure release (venting) system and fire suppression equipment are also evaluated. Class 1.1 and 1.2 materials use tests similar to those reported in Table III and also require venting and fire suppression. The process of choice is made carefully and the tests are designed to prove the merit of the choice both with laboratory and field tests.

Table II. Four Divisions of UNO Class 1 (Explosives)

Hazard Class and Division Designation	Hazard
1.1	Mass Detonating
1.2	Non Mass Detonating
1.3	Mass Fire
1.4	Moderate Fire, No Blast

Safety Classification for Pyrotechnic Materials in a MIGRAD Mixer:

Test results (5,6) for several candidate materials (Table III) are reported which span the range of energetic capacity. Those values which exceed the threshold are highly suspect and have been known to result in serious fires in the past. Mix No. 1, (M49A1, Trip Flare Mixture) is a "safe" mixture that is insensitive to electrical spark, impact, and friction. It does not have a fast burn rate on the Vee Block tester and it has a low pressure-rate-of-rise. Mixture No. 2, (R256 Tracer Mixture) is friction sensitive as indicated by an Eq value of 45 compared to a threshold level of 100 minimum. Mixture No. 3, (I548 Ignition mixture) is both friction and impact sensitive with readings of 66 ft lb^2/sec and 3.75 in. respectively. Mixture No. 4, (40 mm ignition mixture) is friction, impact, and electrostatic sensitive (ESS). It burns rapidly in the Vee Block tester and has a pressure-rate-of-rise exceeding the threshold level of 200 psi maximum. Mixture No. 4 requires more safety constraints in processing than does mixture No. 1 which has no parameters failing the established threshold levels. Before meaningful processing constraints can be established for a pyrotechnic composition, all safety classification tests (Table I) should be conducted to characterize the pyromixture.

Table III. Key Parameters for Safety Evaluation of MIGRAD Mixers

PARAMETER	THRESHOLD LIMITS	PYROMIXTURES (1)	(2)	(3)	(4)
Vee Burn Time-s/cm:	0.06 (Min)	2.67	0.67	1.81	0.05
ESS-Joules:	1.0 (Min)	>50	>50	>50	0.107
Roto-Friction-ft.lb2/s:	100 (Min)	197	45	66	86
Impact Sensitivity-in.:	3.75	10	10	3.75	3.75
Pressure ROR-psi/s:	200 (Max)	17	42.3	38.2	265.8

Mixer Granulator Drier (MIGRAD):

A search was conducted to identify a commercial mixer (7) that would ensure proper mixing, granulation and vacuum drying of pyrotechnic powders and lend itself to restructuring/tailoring for remote accomplishment of project goals. (See Figure 1) The mixer selected was the 30 liter brandy glass shaped "Dry Disperser Mixer/Granulator" made by Baker-Perkins Chemical Machinery Ltd., a British firm. The mixer has two hydraulically driven impellers. Mixing is accomplished by the mixing impeller located in the bottom of the mixer, while granulation is achieved by the granulating impeller, or chopper, located in the side of the mixer. The hydraulic motors drive the impellers in infinitely variable speeds from 0-650 RPM (mixer) and 0-1000 RPM (chopper). The mixing bowl is jacketed to permit cooling or heating and should meet ASME Class VIII, Division I standards (internal working pressure 170 psig). The mixer is equipped with a hydraulically activated discharge valve that allows automatic and remote unloading of the mixer. Further adaptation of the mixer for pyrotechnic applications consisted of adding a mixer extension with entry ports for adding dry raw materials, liquid binders and deluge water. A vent stack was added to vent a possible fire. The mixer was closed by the addition of a 10 psi rated rupture disc between the mixer extension and vent stack. Drying of the pyrotechnic composition is accomplished by circulation of hot water through the mixing bowl jacket while pulling a vacuum on the mixing chamber. Chilled liquid may be circulated through the mixing bowl jacket to keep the contents cool during critical phases of the mixing process.

A fire detection/suppression system was added to the commercial equipment to achieve the fastest possible response time (10-50 ms) in the event of a fire. The fire detection sensors consist of infrared radiation sensor, pressure sensor (8 psi rated) and temperature sensor (210°F/99°C rated) installed directly in the mixer extension. Ultraviolet (UV) radiation sensors monitor the operating bay and the vent stack. Deluge water is delivered independently through a Primac valve with preprimed deluge lines, and through a pressurized water storage reservoir and explosively actuated deluge valve located at the mixer. The fire detection and suppression system is automatically monitored for system faults and controlled by instrumentation provided by Detector Electronics Corporation, Minneapolis, Minnesota.

Steps in the manufacture of a typical batch are reported in Table IV.

Figure 2 graphically represents the blending of dry materials at various mixer and chopper speeds by plotting the product temperature, air temperature above the product bowl, and vacuum versus time. The mixer and chopper were in continuous operation with variable values from the beginning of mixing until approximately 13 minutes into the process. At that time, the mixer and chopper were stopped and pulsed (P) for 3 seconds at 50 RPM at various time intervals.

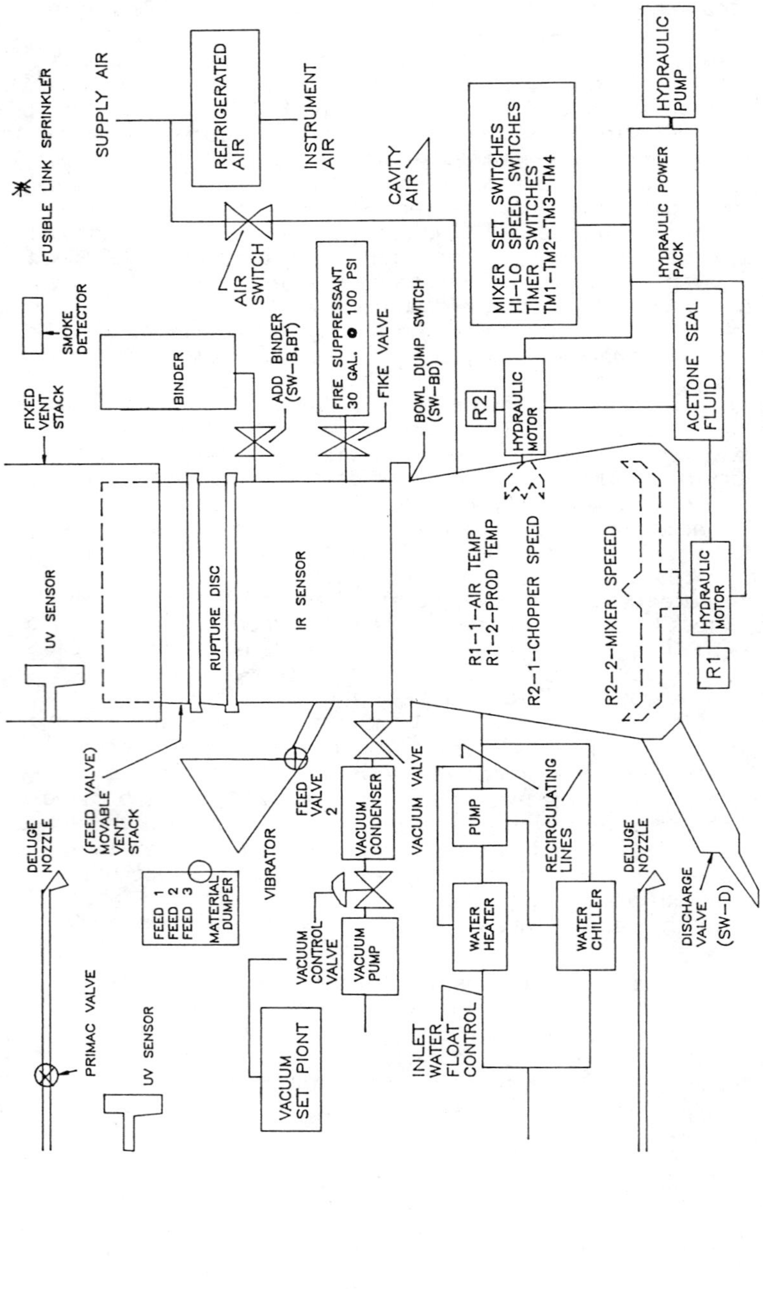

Figure 1. 30 Liter Migrad Mixing System

Table IV. Steps in Remote Manufacture of Starter Mix (Class 1.1), Using a MIGRAD Mixer

1. Place preweighed raw materials in dumpers.
2. Place binder in tank.
3. Remotely load raw materials into mixer (Dumpers place raw materials into feeder hopper).
4. Dry blend the raw materials for 3 minutes (unless safety concerns prohibit dry blending).
5. Add binder and run chopper to achieve granulation.
6. Introduce hot water into mixing bowl jacket and vacuum to the mixing bowl to remove volatile solvents.
7. Control mixing speed, time, temperature, and vacuum until drying is complete.
8. Open discharge valve to discharge mixture into awaiting containers.
9. Clean mixer by flushing with cleaning solution.

By examination of Figure 2 starting at zero time, the dry ingredients were first mixed well. Next the mixer and chopper speeds were reduced while liquid binder was added. After binder addition, the mixer speed was increased and vacuum and heat were applied to the bowl. The decrease in product temperature indicates evaporation of solvent during the drying step. At this point, the mixer and chopper impellers were stopped, other than occasional pulsing to facilitate drying of the mix. When there was a constant temperature differential between the product and the air over the product bowl, the drying was stopped. Typical drying times are 35-45 minutes. When dry the pyromixture was unloaded via a remotely operated, hydraulically driven discharge valve. Particle size and volatiles were determined and the mix was function tested. Test results indicate a homogenous product.

Fluidized Bed Granulator:

The fluidized bed spray granulation process equipment (Glatt Unit) was manufactured by the Glatt Company in West Germany and distributed in the United States by Glatt Air Techniques, Inc., Ramsay, New Jersey. Tests indicated (8) the Glatt WSG-300 unit was capable of mixing, granulating, and drying a 1000 pound batch of M18 colored smoke mix. PBA has two such units. Each unit (Figure 3) consists of a stainless steel product container (bowl), product dolly, lower support section, and upper mix chamber section. The product bowl may be removed from the stationary unit on the product dolly. The mix, or expansion, chamber section of the stationary unit contains the binder spray nozzle. The liquid binder is

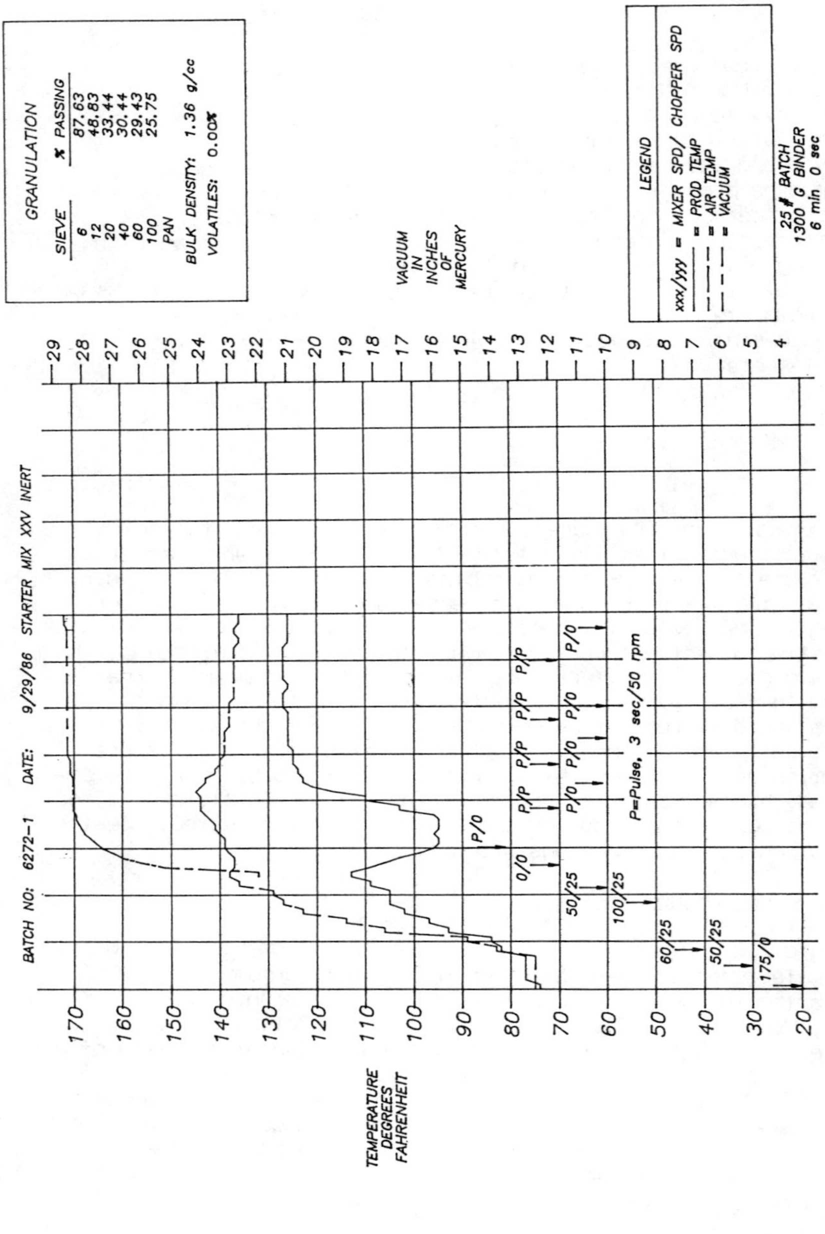

Figure 2. MIGRAD Operating Parameters

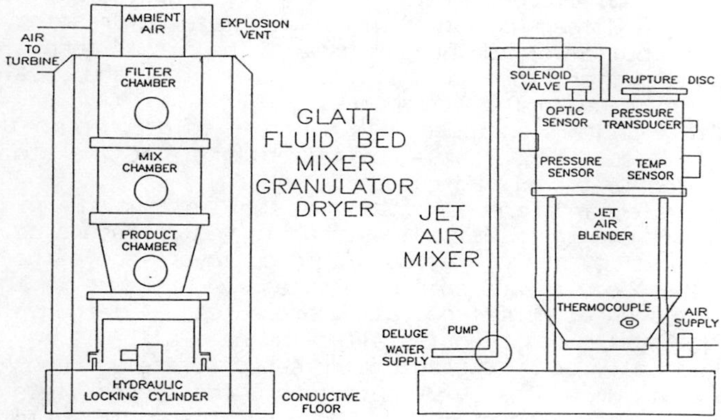

Figure 3. Glatt and Jet Air Mixers

atomized using pressurized air. The mix chamber can be removed when necessary to facilitate filter cleaning. On top of the mix chamber is a conductive cloth filter attached to a shaker arm which is used to prevent loss of materials by returning fines to the fluidized bed at regular intervals. The outlet air flap is also located in this chamber.

The inlet air duct, starting at the roof level of the cubicle, leads to the inlet air flap, inlet air filters, and steam heated heat exchanger coils to heat the inlet air. This heated air is directed through a 1/2" supporting grid and 100 mesh screen on the bottom of the product container.

The equipment operates by negative pressure differential generated by a turbine. Air is drawn in the inlet duct, throttled at the inlet flap, cleaned at the inlet filter, heated at the steam heated coils, and then used to fluidize the material resting on the screen in the bottom of the product bowl. The air is then cleaned at the outlet filter, throttled at the outlet flap, and accelerated through the turbine to the atmosphere.

Each Glatt unit is equipped with a binder solution pump that supplies the liquid binder at a controllable rate to the atomizing nozzle located in the mixing chamber.

Various safeguards are designed into the Glatt unit. In case of material ignition, the Glatt is designed to vent pressure buildups through the roof by the operation of hinged covers above the filter chamber. These covers are opened by excessive pressure in the Glatt and will not open at normal operating pressures. A large volume water deluge system has been installed in the Glatt. It may be activated manually or by two automatic sensor systems. One automatic system operates by UV detection. The sensor views through the window of the mixing chamber. The other automatic system operates by temperature sensing. A thermocouple located above the mixing chamber is activated by a temperature that exceeds 210°F.

Operation of any of the three systems (UV, temperature, or manual control) results in the following actions:

1) The deluge valves in the Glatt and the cubicle sprinklers are activated.
2) A signal is sent to the Fire Station.
3) The Glatt will shut down and the fluidized materials will settle.
4) Building fire alarms will sound.

<u>Fludized Bed Mixer Operation.</u> The Glatt unit is equipped with sensors to monitor the granulation process. The monitors are located on the control panel in the control room. Pressure sensors are installed to monitor the pressure drop across the fluidized bed of material being granulated and across the outlet filters. Temperature sensors monitor the temperature of the incoming heated air and the exiting air. A pitot tube is installed in the inlet air duct to monitor the flow rate of incoming air. These monitors are used in the control of the granulation process.

The Glatt unit is equipped with adjustable valves and timers that are used to control the granulation process. The phases of granulation are: Mixing, Spraying, Drying and Cooling. Timers control the time for the various phases of the granulation process. The inlet air flow is controlled with either the inlet air flow control flap (valve) or the outlet air flow control flap (valve). The heat input into the fluidized bed granulation process is controlled with a steam control valve. The settings for the sensors and controls used in the Glatt unit are listed (Table V).

After the control panel has been preset with the desired control parameters, the colored smoke mix batch is granulated with little operator intervention. In the first phase of granulation, the Mix phase, the raw ingredients are mixed with heated air. The Mix phase is long enough to mix the ingredients but short enough to prevent stratification. When the Mix phase is complete, the Spray phase begins and the binder is sprayed into the fluidized raw ingredients. This granulates the raw ingredients. The spray timer controls the length of time that binder is sprayed. During the Spray phase the fluidized bed becomes heavier due to the addition of the binder and binder solvent even though some solvent (water) is being removed by the heated fluidization air. The fluidization air volume is increased at this point by the opening of the inlet air control flap with a preset timer. When the spray timer runs out, the Spray phase is finished and the Drying phase begins. During the Drying phase the fluidized bed becomes lighter due to the removal of the binder solvent (water). The fluidization air volume must be decreased when the bed becomes lighter to prevent product entrainment into the outlet filters and decrease product attrition. Air volume reduction is achieved by closing the inlet air control flap with a preset timer. When a preset final temperature limit is reached, the Drying phase is complete. The batch may be cooled to the desired temperature or optionally processing may be stopped at this point. For colored smoke mix, processing is stopped before cooling to prevent the particle size attrition that occurs during the Cooling phase.

Binders. Perhaps the most important variables involved in granulation are those related to the binder. Granulation is dependent on not only the type of binder used, but also on the concentration of the binder, the spray rate of the binder, the spray size of the binder, and the total quantity of binder in the mix. Several binder types have been tested. Two binders were found that efficiently granulate the components used in colored smoke mix production. They are polyvinylpyrrolidone, a white, free flowing powder that is soluble in water and organic solvents, and polyvinyl alcohol (PVA), a white to cream colored powder that is water soluble. PVA was found to be the most reliable binder. The concentration of the binder effects granulation because it must be dilute enough to flow but concentrated enough to prevent adding too much solvent to the mix. The binder must be applied slowly enough so that too much solvent is not added to the mix at once, but rapidly enough to limit attrition during the spray phase. Enough binder must be added to granulate the mix. The spray size of the binder is controlled by adjusting the spray atomization air pressure. A high atomization air pressure atomizes the binder. A low atomization air pressure results in some binder droplets. The optimum atomization air pressure results in a spray size which will give a desired particle size distribution in the final mix. The desired particle size distribution is predominately in the 40-100 mesh sieve size. (Figure 4)

Table V. Table of Glatt Process Settings

Acrison Weigh Feeder Set Points

Feed Rate Set Point: 90
Dribble Speed: 5.0
Dribble Point: 98

Glatt Unit Timer Set Points

Mix Time: 3 Minutes
Air Volume Set Point II Time: 15 Minutes
Filter Shake Interval: 45 Seconds
Filter Shake Time: 5 Seconds
Spray Time: 46 Minutes
Air Volume Set Point III Time: 15 Minutes.

Glatt Unit Valve Set Points

Steam Valve Set Point 1: 85
Steam Valve Set point 2: 95
Inlet Air Flap Position: 100
Air Volume Set Point 1: 38.
Air Volume Set Point 2: 44
Air Volume Set Point 3: 38
Atomization Air Pressure Preselection: 2.5 bars
Atomization Air Pressure: 4.0 Bar

Glatt Unit Limit Set Points

Interuption Mixing: $35^{\circ}C$
Operation Cooling: $35^{\circ}C$
Operation Drying: $70^{\circ}C$
Inlet Air Temperature Limit: $90^{\circ}C$
Exhaust Air Temperature Limit (Operation Cooling): $35^{\circ}C$
Exhaust Air Temperature Limit (Operation Drying): $65^{\circ}C$

The binder and fluidization air parameters are in balance with each other and also effect the particle size distribution. If the binder addition rate or binder spray size were increased, the fluidization air temperature or fluidization air rate must be increased to prevent changing the nature of the final product. Likewise if the fluidization air temperature or rate are decreased, the binder addition rate or spray size must be decreased to maintain the same particle size distribution in the final mix.

The binder used in Glatt granulation is a six percent (6%) by weight solution of PVA in water. The binder is accurately weighed and slowly poured into the stirred non-heated water. When all the PVA is added, heat is applied to bring the temperature of the slurry to $185^{\circ}F$. This temperature is maintained for at least 30 minutes, or until all the PVA is in solution. At this point the application of heat is discontinued and the binder is allowed to cool before use.

Full Scale Glatt Mixing. The fluidized bed granulator is one of the most important granulation methods available today, because it combines the unit operations of mixing, granulating, and drying into one system. The fluidized bed mixer, which has long been used by the pharmaceutical industry, fits many pyrotechnic processing requirements such as materials containment, ease of cleaning, homogeneity of mix, and hygiene.

Hazards Analysis-Glatt. Before the Glatt could become a feasible alternative for mixing M18 colored smoke mix, it was necessary to conduct safety classification tests (Table I). The compositions were tested in the Glatt manufacturing process and were found to generate minimal amounts of electrostatic energy during the mixing, granulating, and drying processes. Full scale simulation tests utilizing 740 and 940 pound batches indicated that there were no mass detonation hazards during mixing. Based on the above evidence the Department of Defense Explosives Safety Board allowed an in-process hazards classification of 1.3. This allowed batch sizes to be increased to 1000 pounds.

A critical difference exists between pyrotechnic and pharmaceutical processing in the Glatt granulator. Common pharmaceutical practice involves processing with the operator physically present at the unit to make adjustments as processing dictates. Safety requirements in pyrotechnics processing force remote operation. Since pyrotechnic processing must be performed without the luxury of an operator physically present at the unit to make adjustments as the processing dictates, detailed operating parameters were developed for each pyromixture.

Product Loading. Production of mixes with controlled particle size distributions (Figure 4) can be accomplished in the Glatt, and this control of particle size is essential for successful automated volumetric feeding of a Stokes rotary press. Slug production rates exceed 80 slugs per minute. Therefore a free-flowing product is essential to obtain consistent slug quality. After production, the slugs travel flat on a conductive rubber conveyor (Figure 5) to a gravity track where they are turned on edge. Next they roll down to an automatic slug placement machine. Four slugs are fed into each of two rotary cylinders which rotate the slugs 90 degrees to a vertical position. The eight slugs fall four each into two cans on floating pallets on the conveyor beneath. Proper insertion of the slugs is assured by the passage of rods through the rotary cylinders. The slug filled cans travel to an automated consolidation press where the slugs are consolidated into an integral mass (9,10).

Equipment surveys led to the purchase of a twin feed, 11 station, rotary slugging press (Pennwalt Stokes 523 PBX). The press has a variable production rate of from 60 to 180 slugs per minute. Some features of this press are: double action compression, 30,000 pound capacity, remote pneumatic fill weight adjustment, 7.6 cm maximum slug diameter press, vacuum dust collectors, and explosion proof electrical controls. The press is capable of consistently producing slugs of uniform density and size at a compaction pressure of 5000 lbs.

Variations in the particle size distribution of the smoke mix will occur in any mixing process. Therefore studies were made using three mixes of different particle size distribution from the Glatt process. The particle size distributions were identified as "Dusty"

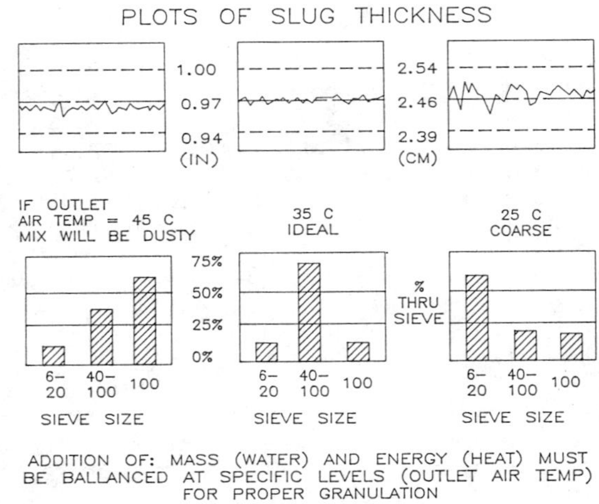

Figure 4. Relationship Between Glatt Operating Parameter, Particle Size and Resultant Slug Thickness

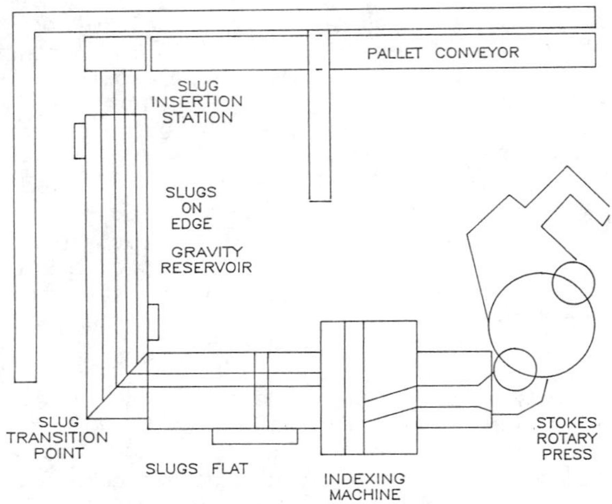

Figure 5. Slug Placement Machine

(many particles smaller than 100 mesh), "Ideal" (most particles 60-100 mesh), and "Coarse" (many particles in 12-40 mesh range). Slugs were produced using the three different mixes with samples taken at one minute intervals for slug thickness checks. Figure 4 shows the general effect of particle size distribution on slug thickness.

As was shown above (Figure 4), a uniform particle size distribution is important to achieve consistent press feeding. A change in particle size distribution changes the rate at which the particles flow and therefore affects slug press loading. Two variables that affect the slug quality are the rate of mix feed into the slug press die and the die fill volume. The mix feed rate varies directly with material hopper discharge height above the feeder. Slug density varies directly with die fill volume. The die fill volume is adjusted by raising or lowering the lower punch on the slug production press. If the mix density changes from "ideal" to "coarse", the hopper is raised to increase the feed rate and the die fill volume is increased to compensate for lower mix bulk density.

Jet Airmix Mixer Smoke Mix Batches.

Hexachloroethane (HC) smoke mix production. Evaluation of the Sprout Waldron 35 cubic foot Jet Airmix Unit (Figure 3) for Mixing 2,200 pounds of white HC smoke mix (consisting of HC, zinc oxide, and aluminum) was conducted (11). The mixer was selected to replace the 840 pound rotary McClelland Blender. Testing revealed that improved mixing was accomplished in about 2 minutes with very few rejected batches. The Jet Airmix unit uses dry, high pressure (250-300 psi) air pulses discharged through angular nozzles to lift, swirl, and blend the material through a tumbling action. Five to twenty short (2-5 sec.) pulses spaced with similarly timed pauses represented a complete mixing cycle.

Safety testing indicated low electrostatic charge generation during mixing. Parametric studies reported the material was difficult to ignite. In-process classification of 1.4 was approved. Four mixers have been in operation at PBA for several years. Loading is from the top, using weigh feeders and air transfer equipment. Mixing air is discharged to a bag house directly above the mixer and then through a HEPA filter. The bag house fines discharge back into the mix and are recycled.

Red Phosphorus smoke mix production. Evaluation of the Sprout Waldron 35 cubic foot Jet Airmix unit for production of Red Phosphorus (RP) M8E1 Smoke Mixtures was conducted (12). Results indicated the mix was stabile and not easily initiated by heat, but sensitive to friction and spark stimuli. The burning time was slow with dense smoke emission.

Full scale mixing studies were conducted without incident using 100, 250, 500, and 1,000 pound batch sizes. Electrostatic charge generation during the blending cycle was several orders of magnitude below that required for initiation.

To further evaluate the mix an electric match was used to initiate the reaction of a 1,000 pound batch of smoke composition. All tests were conducted with the blender equipped with a 16 inch diameter rupture disc rated at 4 psi and an internally mounted UV dectector and water deluge. Without the use of the rapid fire detection and water deluge, a massive "fire ball" was released. With the use of the rapid fire detection and water deluge, there was no mass fire and the mix was dumped into water for continued fire suppression. Any fire with RP results in the formation of white phosphorus (WP). WP must be covered with water since it ignites spontaneously when exposed to air. Processing studies were conducted to determine the best methods for pollution abatement since WP/water mixtures are toxic at 29 ppb for blue gill bream and since high levels of phosphorus (reported as total phosphorus) may not be dumped into the environment.

There was no significant damage to equipment in the fire tests, and it was demonstrated that a Jet Airmix mixer may safely handle the mixing of RP formulations on a routine basis. Since a high risk of fire is always associated with any method of transfer of RP, a pneumatic conveying system (dynamic air, two phase positive pressure transfer system) was evaluated to load RP into the Jet Airmix mixer. Electrostatic charge measurements were minimal and indicated the system was satisfactory to load the blender.

All work was conducted with "oiled" RP as supplied by ERCO Limited, Canada. The "oiled" RP is much less sensitive than "non-oiled" RP.

Conclusion.

Proper safety testing and classification of pyrotechnic energetic capacity will allow the selection of appropriate, remotely operated, commercially available equipment. This equipment can be installed in less costly structures and plant sites for the manufacture of pyrotechnic materials in a safe and economical manner.

Often, considerable problems arise in cost and safety when pyrotechnic formulas are selected from the literature and used without regard for the energetic requirements of the task to be accomplished. For example starter mix formulas mav be too reactive for their intended use, but they could be used if they were modified and tested relative to percent composition, particle size, consolidation pressure, purity, etc. to gain a 1.3 or 1.4 UNO classification. The continued addition of ingredients over the years for heating or cooling of a formulation without regard to the basic chemistry of the mixture was a problem that was noted through review of many formulations in the literature. Thus many examples may be found where "extra" ingredients have been included which tend to negate each other and raise production costs.

The American Pyrotechnics Association, P.O. Box 213, Chestertown, Maryland 21620, an industry association, provides assistance to manufacturers that require more information. Annual Summer Symposia in Pyrotechnic Chemistry are also offered by Washington College, Chestertown, Maryland 21620. The International Pyrotechnics Seminar on Explosives and Pyrotechnics is offered on a biennial basis. Additional information on these seminars may be obtained from IIT Research Institute, Chicago, Illinois 60616.

Literature Cited

1. McLain, Joseph H. Pyrotechnics; The Franklin Institute Press Philadelphia, Pennsylvania. (1980).
2. Conkling, John A. Chemistry of Pyrotechnics; Marcel Dekker, Inc. New York (1985).
3. Meyer, Rudolf. Explosives; Verlag Chemie, New York (1977).
4. Safety Regulation (AMC-R 385-100, Chapter 17), U.S. Army Material Command, Alexandria, Virginia 22333-0001 (1985).
5. Armour, Carl; and Smith, Lloyd A.; "The Invention of a New Type of Friction Sensitivity Apparatus," RDTR No. 60, 11 June 1965, U.S. Naval Ammunition Depot, Crane, IN
6. McIntyre, F.L. Safety Enhancement Program for MIGRAD Mixer Study; U.S. Army Armament, Munitions, and Chemical Command, Aberdeen Proving Ground, Maryland, 21010 (1985).
7. Aikman, Loy; Shook, Thomas E.; Lehr, Robert; Robinson, Eddie; and McIntyre, F.L.; Improved Mixing, Granulation and Drying of Highly Energetic Pyromixtures; Pine Bluff Arsenal, Pine Bluff, Arkansas 71602-9500 (1986).
8. Garcia, David; Aikman, Loy; Ables, Larry; McIntyre, Fred; and Shook, Thomas; Smoke Mix Facility (Glatt); Pine Bluff Arsenal, Arkansas 71611 (1983).
9. Garcia, David J.; Aikman, Loy M.; McIntyre, F.L.; and Shook, Thomas E.; Rapid Plant Scale Mixing Granulation and Loading of Dry Materials During the Manufacture of Colored Smoke Munitions; Pine Bluff Arsenal, Pine Bluff, Arkansas 71602-9500 (1981).
10. Aikman, Loy; Garcia, David; and Shook, Thomas E.; New Fill and Press Technology for Production of Colored Smoke Grenades; Pine Bluff Arsenal, Pine Bluff, Arkansas 71602-9500 (1982).
11. Fortner, Wendell; Yeidell, Steven L; and Shook, Thomas E.; HC Product Improvement Studies; Pine Bluff Arsenal, Pine Bluff, Arkansas 71602-9500 (1977).
12. McIntyre, Fred; Amend, R.J.; and Smith, M.; Blending Technology for Red Phosphorus Smoke Compositions, U.S. Army Armament, Munition and Chemical Command. Aberdeen Proving Ground, Maryland 21010-5423 (1985).

RECEIVED May 1, 1987

Chapter 10

Engineering Design for White Phosphorus Filling Operations and Facilities

Harold D. McKinney

Pine Bluff Arsenal, Pine Bluff, AR 71602-9500

This paper describes the development of a system and facilities for safe, efficient, and accurate filling of white phosphorus (WP) munitions. This new development replaces dip-fill operations used by the U.S. Army for over thirty years, a production method that was hazardous to operating personnel and generated unacceptable quantities of phosphorus contaminated water and gas. The new development, Volumetric Filling, is relatively pollution free and exceeds the U.S. Army's standards for filling of white phosphorus munitions.

Since World War II, Pine Bluff Arsenal has produced millions of white phosphorus (WP) munitions for the United States Department of Defense. White phosphorus has a specific gravity of 1.728 at 145°F (the temperature that is normally used for WP filling operations) and melts at 111.4^{o}F; it ignites spontaneously in atmospheric air and generates a dense white smoke, phosphorus pentoxide (P_2O_5). Phosphorus pentoxide reacts with moisture in the air to form phosphoric acid. WP munitions were used by U.S. military forces and their allies to mark targets and to provide smoke screen coverage for troops and equipment in combat zones. These munitions were produced primarily by the dip-fill or wet-fill method illustrated by Figure 1. The method is called dip-fill because empty munition bodies are dipped below the molten phosphorus level in an open tank until the munitions are filled with liquid phosphorus. The method is also called wet-fill because a water overlay is maintained over the liquid phosphorus (in the fill tank) to prevent spontaneous combustion of the chemical element and because the filled munition will have a slight water overlay (up to 1/8" column height allowed).

Contamination of line equipment on a dip-fill line is a constant problem. During filling operations, WP contamination is transferred from filled munitions and pallets to surfaces of accessory equipment until the filled munition enters the cleaning station. Large quantities of water and gas are contaminated from:

This chapter not subject to U.S. copyright
Published 1987 American Chemical Society

a. partial aspiration of filled munitions to the correct height of fill

b. necessary fire control action of spraying water on munitions and pallets on the filling line

c. munition cleaning station

Because of the disadvantages of the dip-fill method, the U.S. Army began efforts to provide a more acceptable method to fill and close white phosphorus munitions. This effort included design and installation of a small prototype WP "Height of Fill" (HF) production line at Rocky Mountain Arsenal for 105mm, M60 rounds, and fabrication and test of a two nozzle HF line at Edgewood Arsenal. After the Edgewood Arsenal HF line was successfully demonstrated in Maryland, the equipment was moved and reinstalled at Pine Bluff Arsenal where approximately 30,000, 2.75", MK67 WP rockets were filled for the U.S. Navy and 750, 175mm, XM510 WP rounds for the U.S. Army.

After satisfactory operations at Pine Bluff Arsenal, Edgewood Arsenal prepared a total design package for a prototype WP multi-munition HF production line. The specified production rate was 8,000 munitions per eight hour shift. A contract (Project No. 5680242) to design, fabricate, install, and de-bug the system at Pine Bluff Arsenal was awarded in June of 1969. The new HF line was installed at Pine Bluff Arsenal in 1971. After numerous attempts to operate the facility ended in failure, the contract was terminated in late 1972. Serious problems with the HF filling system was the primary reason for failure of the new production line. Shortly after termination of the contract, Pine Bluff Arsenal conceived and developed a "volumetric dry fill" concept that proved to be an outstanding method for production of WP munitions.

A project (No. 5751274) was approved and funded by the Army's Production Base Modernization and Expansion Project Management Office to prove out the Pine Bluff Arsenal volumetric filling concept on a production basis.

Some of the contractor-furnished equipment for the original dry fill production line (conveyors, munition pallets, filling station framework and fill tank, hydraulic units and electrical power circuits) was modified and used during early development work.

Description of the Pine Bluff Arsenal Volumetric Filling Concept

The Pine Bluff Arsenal white phosphorus volumetric filling system (U.S. Patents 4,002,268, 11 January 1976, and 4,043,490, dated 23 August 1977) was conceived and developed by Pine Bluff Arsenal in 1973 and has been used in filling WP munitions since early 1974. This development has provided a safe, clean and efficient method for processing WP munitions (30% reduction in manpower requirements and a 90% reduction in air and water pollution). The system is an extremely accurate production filling method. This accuracy is very important in WP operations since any adjustment in munition volume is hazardous and inefficient. The line changeover from one munition to another is accomplished by two experienced men in one day.

The filling method (See Figure 2) is essentially a fail-safe system in that controls are designed to prevent double-cycling. The filling valve and the reservoir valve are electrically interlocked so

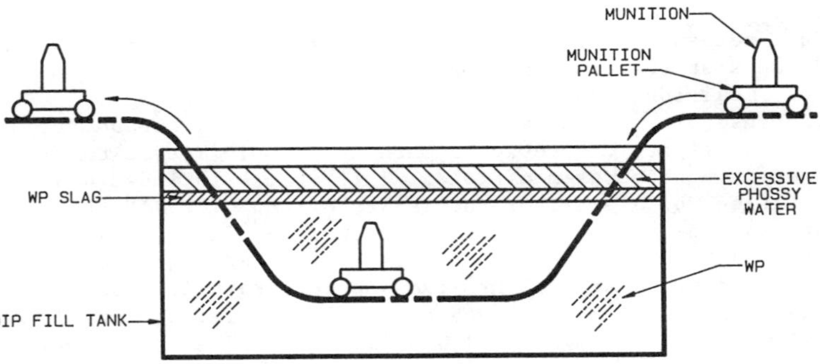

Figure 1. WP munitions dip fill schematic.

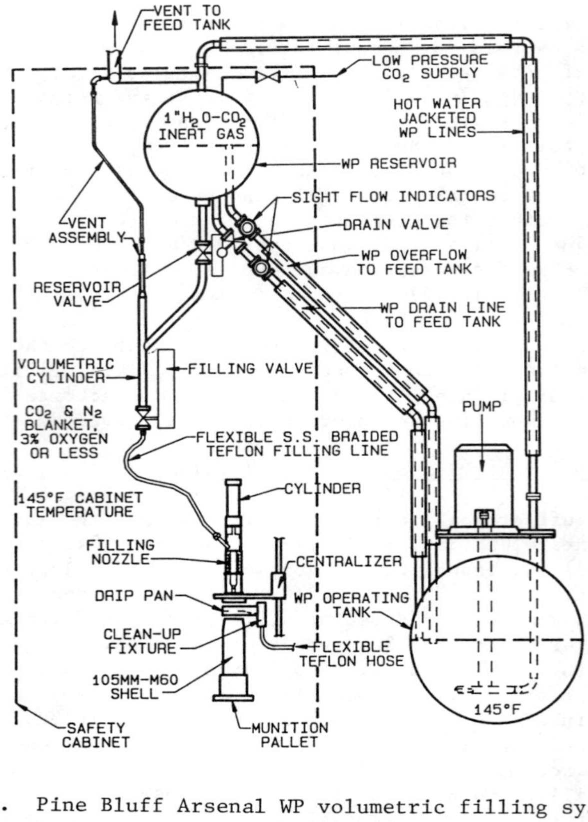

Figure 2. Pine Bluff Arsenal WP volumetric filling system.

that only one of the two valves can be open at any given time and the other must be completely closed and remain so until the other valve closes. This direct control feature prevents an operator error "spill" from occurring in the fill station. Also, all WP filling, reservoir, and control valves are pneumatically-operated (spring-return closure), fire-safe ball valves that close immediately upon interruption of electrical power or air supply.

The automatic filling cycle begins when an empty munition moves into position under a filling nozzle (the volumetric chamber has been previously charged). The filling nozzle (See Figure 3) is inserted into the munition (the nozzle spring is compressed and opens the fill port) and the filling valve opens for a timed interval, dispensing a fixed, repeatable volume into each munition presented. After the filling time is terminated, the filling valve closes, the nozzle retracts (the nozzle spring expands and closes the fill port) and the reservoir valve (See Figure 4) is opened for a timed interval, allowing molten WP to flow from the WP reservoir tank through the reservoir valve into the volumetric chamber, rising until the bottom of the adjustable vent tube is covered by WP. At this time, the gas trapped in the volumetric chamber is slightly compressed. The molten WP then flows through the path of least resistance, which is through the adjustable vent tube. WP flows through the vent tube until the liquid height in the vent tube is equal to that in the reservoir tank. The reservoir valve closes, and a preset and repeatable volume of WP is ready for dispensing into the next munition presented. The filling volume can be changed by a simple adjustment of the vent tube. The volume is decreased when the bottom of the vent tube penetrates further into the volumetric chamber and is increased when it is raised to a higher level in the chamber. Figure 4 shows the original volumetric chamber used under Phase I of this development and an improved volume chamber that was used for Phase II work. Figure 5 shows the accuracy of the volumetric chamber used during Phase I of this development. The improved chamber provided increased accuracy (See Figure 6) required for smaller munitions such as the 60mm M302. The small diameter in the vent tube adjustment area prevents any serious volume variations caused by changes in gas compression.

Design Considerations for Development of the Pine Bluff Arsenal Automatic Volumetric WP Filling Facility

Design Considerations. For this new facility requirements were set as follows:

a. Meet or exceed the filling accuracy requirements for all WP munitions filled by the U.S. Army.

b. Have the capability to fill all WP munition bodies up to 18" in height.

c. Provide safe working conditions for operators.

d. Reduce manpower requirements and increase efficiency.

e. Production rate of 24 munitions per minute.

f. Reduce significantly air and water pollution associated with dip-fill operations.

g. Provide fail-safe, automatic operation where possible.

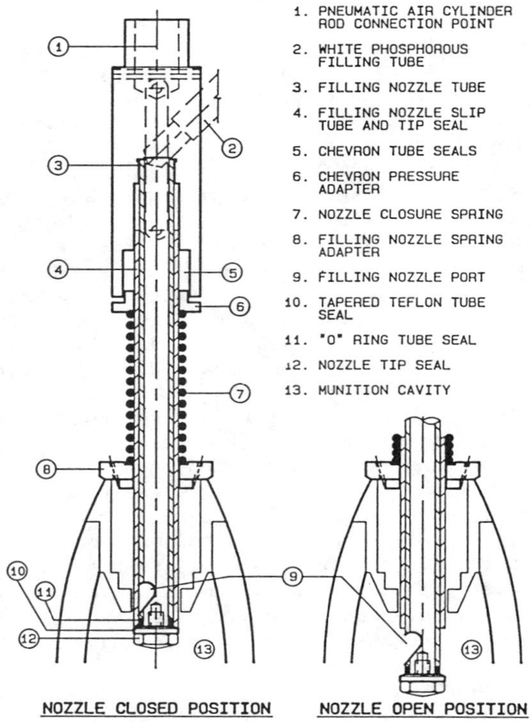

Figure 3. Pine Bluff Arsenal WP filling nozzle.

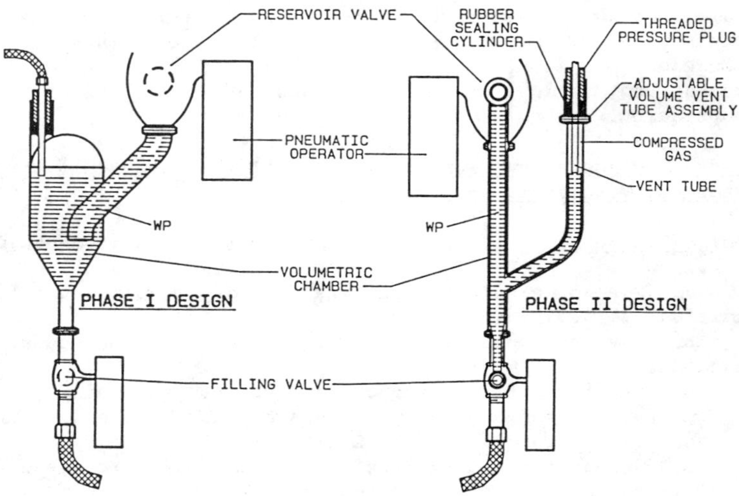

Figure 4. Volumetric cylinder development.

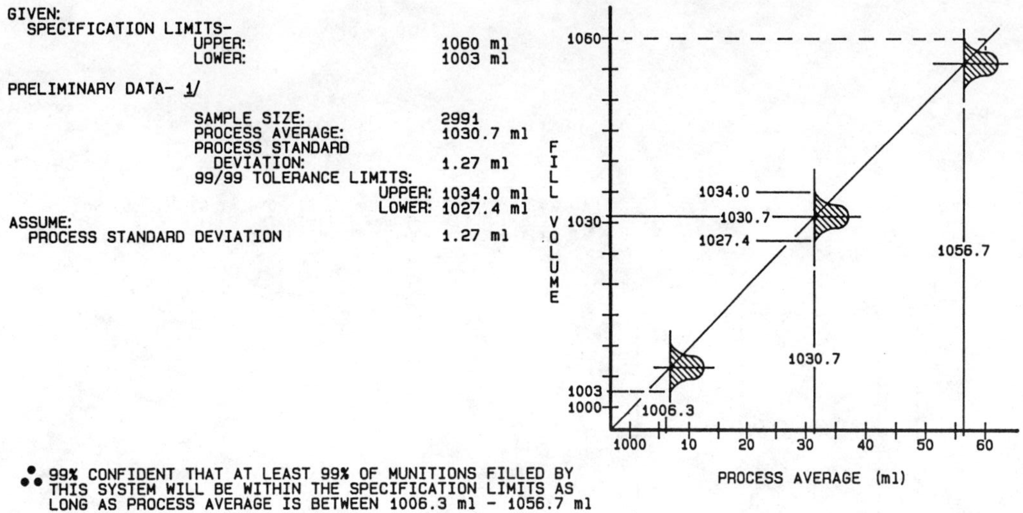

Figure 5. Pine Bluff Arsenal phase 1 WP volumetric filling accuracy.

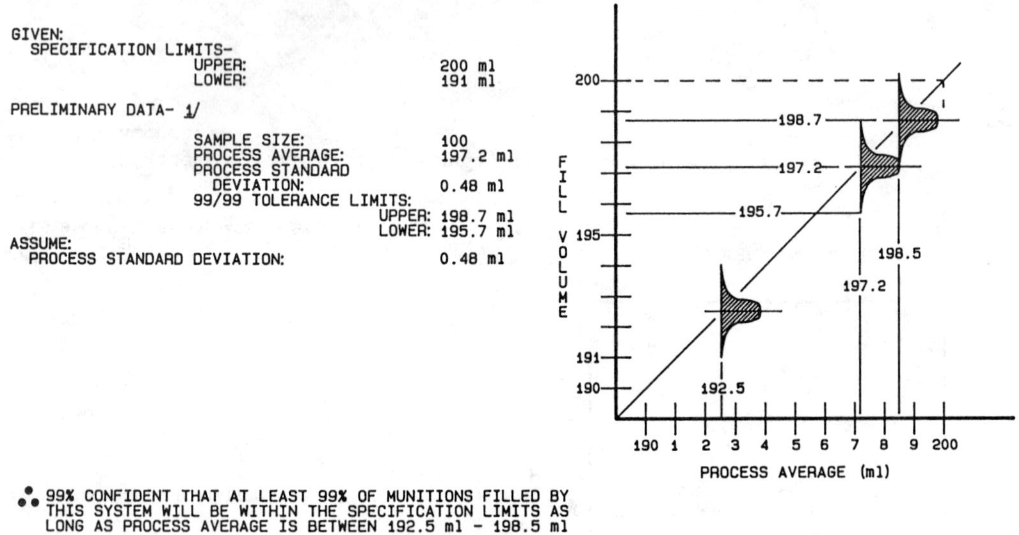

Figure 6. Pine Bluff Arsenal phase 2 WP volumetric filling accuracy.

The major effort for the Pine Bluff Arsenal development work was concentrated on the filling system with other work stations receiving as much attention as time and funding allowed. The facility is listed as WP Line No. 3, Building 34-110, at Pine Bluff Arsenal. A partial listing of major material and equipment requirements and components for the facility is as follows:

Materials Specifications. All piping, valves, tanks and other metal parts which are in direct contact with WP are constructed of 316 low-carbon (L), stainless steel (SS) for welded connections and 316 SS for screwed connections. Hot water jackets are fabricated from Schedule 40 black iron pipe. All WP flexible filling, drain, and vent lines are fabricated from SS braided Teflon hoses. All automatic valves for WP service, including filling and reservoir valves, are pneumatically-operated, fail-safe, spring-return, fire-safe ball valves. These automatic valve units include a waterproof limit switch package that contains two double-pole, double-throw switches. All pipe fittings other than welded joints are made with quick-clamp compression-type fittings with Teflon gaskets.

Automatic White Phosphorus Filling Station. The filling station has eight complete filling units and consists of the following items: (See Figure 2)

a. Pallet stop systems with alignment shot pins for accurate alignment of a munition and pallet under the filling nozzle.

b. An automatically-operated drip pan (for all eight filling units) that retracts when a munition is in filling position and extends horizontally after filling is completed and the filling nozzles are retracted in the vertical plane.

c. A filling nozzle with guide system for accurate alignment with munition-filling openings. The munition-filling nozzle (illustrated by Figure 3) is spring loaded with Teflon chevron seals in the body between moving parts to prevent external contamination of metal parts of the nozzle. A nozzle tip seal (including an "O" ring and a Teflon tip seal) reduces drippage of WP after the nozzle closes. The nozzle serves as a valve with the primary function to reduce drippage after each filling operation. The nozzle is moved in the vertical direction (See Figure 2) by a pneumatically-operated cylinder. The filling nozzle is connected to the filling valve by a flexible filling hose. The framework on which the nozzle, cylinder, and alignment guide are mounted is adjustable in the vertical direction in order to accommodate large or small munitions.

A clean-up fixture is furnished to drain WP from above the reservoir valve after the filling tank has been drained of WP and replaced with phossy water. The clean-up fixture is used at the end of a shift and prior to start up. At the end of the shift (after the fill tank has been drained of WP and replaced with phossy water), the WP above the filling valve is flushed through the filling system into the clean-up fixture and back to the WP operating tank; the filling system is then operated through several cycles to clean the reservoir and filling valves, volumetric chamber, vent assembly, filling nozzles, and filling hoses. The system is then secured at the filling station.

The clean-up fixture is part of the drip pan. The clean-up fixture and the drip pan are drained by gravity to the WP operating tank. The fixture is connected by a hose to a drain pipe that connects to the WP operating tank. The fixture is operated in the horizontal plane by a cushioned-stroke pneumatic cylinder for smooth operation. The fixture travel is two-position travel. The first travel or shorter distance is for drip pan function during normal operation and the greater or over travel is for clean-up operations. The clean-up fixture opening is such that the filling nozzle seats on the clean-up fixture as it would on a munition.

Filling Conveyor. Filling line transfer system is an automatic, nonsynchronus, variable-speed drive unit complete with filling pallets and nests for the five different munitions filled and closed on the line. This unique system moves work pallets from station to station and provides accurate shot pin alignment for the work piece as various operations are performed, and features automatic acceleration and deceleration of the filling pallets. Operation is unusually smooth, quiet and safe.

Inert Gas Cabinet System. The cabinet system encloses an automatic WP filling system and a weighing station and contains an atmosphere that is maintained at 3% O_2 or less to reduce the occurrence of smoke generation or fire should any WP become exposed inside the cabinet area. The cabinet has a temperature controlled steam heating system that maintains the cabinet space at 145°F, and an inert gas distribution system for maintaining the inert (CO_2 and N_2) atmosphere, and an entry and exit air lock to reduce the inflow of air during filling operations. Flexible rubber strips are used at the air lock sites. Small exhaust fans (150 CFM) are used at the air lock locations to prevent inert gas from discharging to the work area. The exhaust fans are vented to the outside atmosphere.

The cabinet enclosure has hot water wash hoses with nozzles for any clean-up necessary, and contains both a manual and an automatic fire control system. Lexan plexiglass doors and windows are provided at the front or operating face of the enclosure for observation and access. All doors and enclosures are essentially air tight. Adequate lighting is provided for the interior of the housing.

Sequence of Operation of the Pine Bluff Arsenal WP Volumetric Filling Line.

An operator (See Figure 7) lifts empty munitions from a standard wooden pallet (elevated for operator access) and places munitions into the filling line pallets. The pallets are released automatically when the munitions are dropped into pallets. The items are conveyed to an empty munition weigh station and are weighed simultaneously and the weights recorded in the programmable logic controller (PLC) for later use in matching and interface with the data from a final weighing of the munitions after the rounds have been filled.

After weighing, the munitions are released to a four-unit vacuum purge station. Automatic vacuum/purge nozzles make a vacuum/pressure seal on top of the empty munitions, and a three-way automatic valve

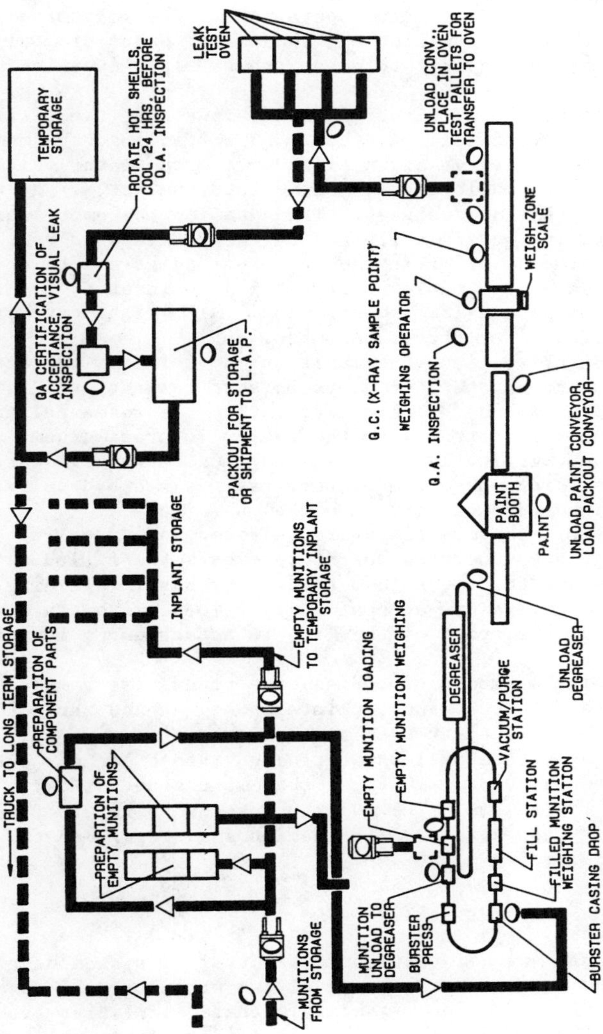

Figure 7. Schematic flow diagram of Pine Bluff Arsenal WP volumetric filling facility.

opens and connects the air-filled empty munitions to a vacuum surge tank. The empty munitions are evacuated to 29" Hg. The three-way valve then closes the vacuum port and opens up to an inert gas port which breaks the vacuum in munition cavities with low pressure inert gas (CO_2 and N_2). This station reduces the amount of burning and smoke generation during WP filling operations. The vacuum/purge nozzle retracts from the munition and the pallet stops drop allowing the four munitions and pallets to move into an eight-munition accumulator.

From this accumulator, eight munitions move into the filling station containing eight filling heads. After filling, the munitions leave the station and arrive at an eight-unit accumulator which releases four pallets/munitions at a time into the net weight station where the filled items are weighed. The data for the empty weight in memory and the filled weight collected by the PLC is used by the PLC to calculate the amount of WP in the munition bodies and to determine if the munitions are acceptable or reject and to identify the reject munitions for removal and later correction. All weight data is automatically printed for record purposes.

From the weight station, the munitions move into a burster casing station where an operator drops burster casings into the filled and accepted rounds. The operator then presses a release button and the munitions travel to the hydraulic press accumulator. The accumulator automatically releases four munitions with bursters into the press station where the burster casings are hydraulically pressed (metal interference fit) into the munition.

After pressing, the munitions are released and travel to a manually-operated stop where an operator removes the filled and closed munitions and transfers them to the degreaser (cleaning) unit. The empty pallets are automatically released from this station and travel back to the front of the line to accept empty munitions for another cycle.

After cleaning of munition bodies, the rounds are sampled for Quality Assurance lot acceptance, painted, weighed and zoned (if required), and then placed in oven test pallets.

The filled oven test pallets are loaded into a hot air oven and the munitions are heated to 210°F and then maintained at that temperature for 15 minutes. The munitions are then returned to the WP plant for leak inspection, palletization and storage or transfer to an ammunition loading plant.

Status and Plans for WP Operations

The Pine Bluff Arsenal volumetric WP dry fill system development work has resulted in the installation of two production lines and one small experimental production facility. These facilities are used to produce WP bulk-filled munitions, wick loaded canisters, and experimental munitions. Figure 8 is a photograph of the original single-station prototype filling station used to prove out the Pine Bluff Arsenal volumetric concept. Figure 9 shows the first production line fabricated and operated at Pine Bluff Arsenal using the concepts proven on the prototype unit. Figure 10 is a photograph of the filling station of the production line for 155mm, M825 wick-type munitions. This facility uses the Pine Bluff Arsenal concept

Figure 8. Original Pine Bluff Arsenal WP prototype filling station. Photo courtesy of the U.S. Army.

Figure 9. Front view of the first WP production facility using the PBA volumetric filling concept. Photo courtesy of the U.S. Army.

Figure 10. Front view of the second WP production facility (PBA concept with vacuum assist) for filling of the 155mm, M825 wick canister. Photo courtesy of the U.S. Army.

Figure 11. Most recent facility using the PBA concept for limited production/experimental filling of standard and new munitions and canisters. Photo courtesy of the U.S. Army.

but differs in that vacuum is used in the filling cycle. Figure 11 is a photograph of the most recent WP dry fill line installed at Pine Bluff Arsenal (1986). The capacity of this small facility is only six munitions per minute; however, the purpose of this experimental unit is to provide limited production of bulk-filled or wicked-type munitions and canisters, and fast set up for filling of new experimental WP items.

Future plans for our WP operations include the replacement of two remaining WP dip-fill production lines with the more accurate, efficient, and safer volumetric filling method described in this paper.

Acknowledgments:

Illustrations provided by James Palmer, Pine Bluff Arsenal.

The contributions of several Government organizations and employees contributed to the success of the Phase I and II development of the Pine Bluff Arsenal volumetric filling production line. Noteworthy support was provided by the following individuals:

Pine Bluff Arsenal:

B.T. Armstrong - Senior Electronic Engineer
Will Bradford - Senior Project Engineer/120mm Wick Round
Bill Miller - Senior Electronic Engineer
Lonnie Witham - Project Engineer/155mm M825 Wick Round
Jackie Smedley - Electronic Technician
Larry Davenport - Mechanical Technician
Earle Mustachia - Mechanical Technician
Tom Woolley - Senior Tool & Die Maker
Elmer O. Woods - Equipment Mechanic
Hubert Gates - Welder
O.B. Summerford - WP Pumper Foreman
William A. Cook - WP Pumper

Edgewood Arsenal:

Frank Stewart - Phase I Project Manager
Merlin Erickson - Phase II Project Manager

U.S. Patents:

1. McKinney, H.D. U.S. Patent 4 002 268 - 1976
2. McKinney, H.D. U.S. Patent 4 043 490 - 1977

RECEIVED April 23, 1987

Chapter 11

Design and Use of High-Speed Detection Systems for Explosives Operations

Kenneth M. Klapmeier and Bernhard G. Stinger

Detector Electronics Corporation, 6901 West 100th Street, Minneapolis, MN 55438

A properly designed fire or explosion suppression system can provide satisfactory protection for applications involving the presence of explosive materials by responding to an ignition source in a matter of milliseconds. In order to be effective, only the fastest equipment and techniques are adequate. A successful system includes an optical detector that responds to the electromagnetic radiation produced by a flame. The detector generates a signal that is used to open a high speed electrically actuated valve. Opening the valve initiates immediate flow of water through the nozzles of a carefully designed piping system to extinguish or contain the fire or explosion.

When considering the use of equipment for detecting and suppressing fires and explosions, munitions manufacturing processes are among the most hazardous. In these applications, little time is available for the system to respond. A reaction time that is only a few milliseconds too slow could result in extensive property damage and even loss of life.

By combining radiation detectors with an ultra high speed water deluge system, response times that are short enough to prevent a catastrophe can be achieved. The high speed deluge system is designed to detect a flame or ignition source and respond by applying large volumes of water in an extremely short period of time (milliseconds). The system consists of the following basic components:

- Flame detectors
- Controllers
- Source of water
- Valve (squib or solenoid operated)
- Piping system with nozzles.

The flame detector is an optical device that responds to the radiant energy that is given off by a flame. When a flame or explosion occurs within the field of view of the detector, the resulting electromagnetic radiation travels toward the detector at

0097-6156/87/0345-0183$06.00/0
© 1987 American Chemical Society

the speed of light. The detector responds to the radiant energy in milliseconds, sending a fire signal to the controller, which in turn generates the signal that opens the valve. When the valve opens, line water pressure is applied to the priming water that is in the pipe behind the nozzles. This causes water to flow from the nozzles, extinguishing the fire. Simultaneously, the controller sends alarm signals to audibly and/or visually indicate a fire occurrence and shut down the associated process equipment.

Ultraviolet Detectors

The ultraviolet (UV) detector (see Figure 1) consists of a gas-filled cold cathode sensor tube that is mounted inside an explosion proof housing. The sensor tube is designed to respond to a narrow band of radiation typically between 1850 and 2450 angstroms. Figure 2 illustrates the general relationship between solar radiation at the surface of the earth and the spectral response region of a typical gas-filled ultraviolet sensor. As you will note, the solar radiation spectrum extends approximately from 2850 to 30,000 angstroms. Therefore, the sensor tube does not respond to solar radiation or normal ambient light.

Radiation is not emitted continuously, but is emitted in small bundles called photons. The energy of a photon is dependent on the wavelength of the radiation. When a photon of radiation is absorbed into a metal such as the cathode (negative plate) of the UV tube, the energy of the photon is imparted to an electron within the metal, causing it to leave the surface of the metal and be drawn toward the anode (positive plate). The energy that the electron must have to leave the metal is called the work function of the metal. The sensitivity range of the radiation detector is dependent upon the work function of the metal used in the cathode.

The sensor tube is filled with an ionizable gas, such that when an electron is emitted from the cathode and is rapidly drawn to the anode as shown in Figure 3, it strikes a gas molecule with enough energy to cause electrons to be emitted from the gas molecule. These electrons strike other gas molecules releasing more electrons. The total number of electrons generated in this manner is typically several million times more than were emitted from the cathode. This current of electron flow is known as the avalanche effect.

The current can be stopped by reducing the applied voltage to the tube so that the emitted electron does not have sufficient energy to cause other electrons to be emitted when it collides with gas molecules.

In a typical UV detector, the current is allowed to flow for a very short period of time before the voltage is reduced and the current stopped. Thus the output of the sensor tube is a series of voltage pulses, the frequency of which is proportional to the intensity of the UV sensed by the detector. The closer a fire is to the detector, the higher the output frequency, and the smaller the flame size that is needed to actuate the system.

In the past, the circuitry in the controller that was used for counting the voltage pulses would amplify and square the pulses, and then use the pulses to charge a capacitor. When the capacitor was charged to a pre-calibrated threshold voltage, the controller

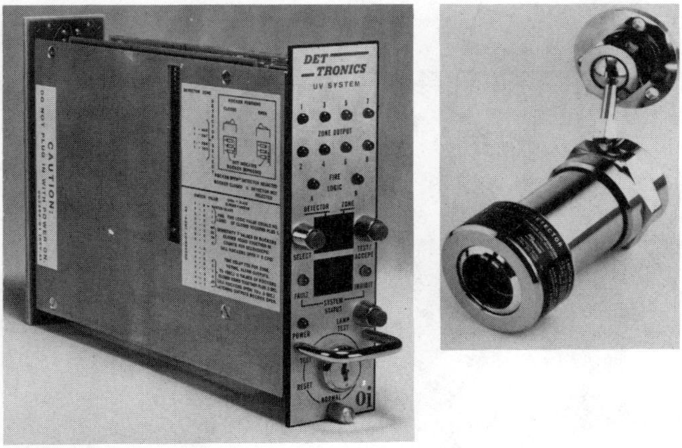

Figure 1. UV Detector and Controller

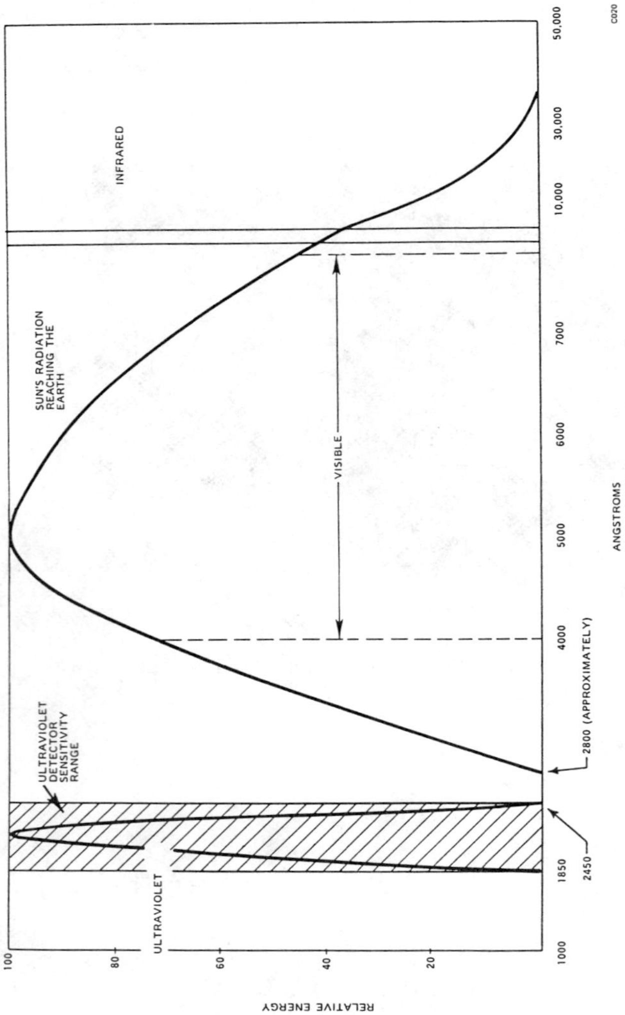

Figure 2. Response Range of Typical UV Detector

generated an output signal that energized the alarm relays and deluge systems.

The use of microprocessors now makes it possible to count and process the digital pulses from the UV detectors. Pulses no longer need to be stored in capacitors, but can be individually counted, entered into the registers of the microprocessor, stored in memory and manipulated like any type of data processing information. This allows the design of flexible ultraviolet fire detectors using programmable memories and switches to provide an infinite number of combinations. Thus we now have a marriage of extremely high gain gas-filled vacuum tube UV detection devices that have existed for many years with state-of-the-art microprocessors. Since the UV detector requires no signal processing other than comparing the radiation level to a preset threshold, a very fast response time is achieved.

Applications. Ultraviolet detectors are ideally suited for applications where rapidly developing fire can occur in a relatively open area. UV detectors can be used to monitor ammunition assembly lines, gunpowder troughs, or open areas that are stocked with hazardous materials. These detectors are not typically affected by extremes of temperature or pressure, adverse weather conditions, high humidity, nor are they sensitive to solar radiation.

In a typical application, UV detectors are used in general or spot coverage locations. General coverage detectors are usually mounted in the corners and along the walls of a hazardous area. They are normally positioned for overlapping fields of view. Their purpose is to detect a fire that occurs anywhere within the hazardous area.

Spot coverage detectors are normally mounted as close as possible to the point of potential ignition. Examples are the extruder/cutter in a high explosives machining operation or the compression point in a shell loading machine. Spot detectors assure the fastest possible detection time by physically being mounted the closest to the point of ignition.

Limitations. Although UV fire detectors have many advantages, they also have their limitations. They will respond to radiation sources besides fire such as lightning or electric arc welding, as well as x- and gamma rays. In some applications, the system may have to be shut down to prevent false alarms when these sources of interference are present. In applications where the presence of x- and gamma radiation is a continuous problem the use of a special nuclear surveillance system is recommended. This system uses dual detectors. One responds to both nuclear radiation and UV from fire. The other is blinded to UV produced by fire and detects only nuclear radiation. The microprocessor based controller uses a special program that utilizes a "count subtraction" technique. By subtracting the output count of the detector that sees only nuclear radiation from the count of the other detector, reliable protection can be assured in applications that normally would be difficult or impossible to supervise. It must be noted, however, that the additional signal processing that is required will increase the response time of the detection system.

It must also be noted that since the ultraviolet detector is an optical device, objects that are able to block its view cannot be allowed to come between the detector and the area to be protected. In addition, smoke and various vapors can significantly absorb UV, making it difficult or impossible for the detector to "see" a fire. It is recommended that the detectors be positioned so that any point within the area to be protected is covered by more than one detector. This will assure reliable protection if a given detector should fail or if its view is suddenly blocked.

Self-checking Feature. UV absorbing contaminants that are present in the environment can accumulate on the optical surfaces of the detector. An accumulation of certain materials, sometimes barely visible to the naked eye, can cause a significant reduction in the level of UV that reaches the sensor tube of the detector. This could make the detector nearly "blind" to UV radiation. An electronic self-testing feature, known as Automatic Optical Integrity, has been designed to guard against such an occurrence. The system generates a calibrated UV test beam from a small tube that is located inside the detector housing beside the UV sensor tube. The test beam passes outside the viewing window of the detector and is then reflected back through the window and into the UV sensor. See Figure 4. The sensor tube then generates an output signal that is sent to the controller, where the intensity is evaluated to determine the relative cleanliness of the viewing window. The test signal does not interfere with the normal functioning of the detector, since it is considerably weaker than a UV fire signal. Therefore, no danger of a false alarm exists. In addition, if a fire should occur during an Optical Integrity test, a fire signal will immediately be generated. The system continuously checks the optical surfaces, electronic components, and inter-connecting wiring of the detector. Any malfunction is detected in a matter of seconds. The controller responds by registering a fault output to alert personnel that a problem has occurred.

When properly applied, ultraviolet detectors can serve as excellent fire detectors in munitions manufacturing. Detection times as fast as 10 milliseconds can be achieved while effectively resisting false alarms.

Infrared Detectors

The infrared (IR) detector is an extremely fast device that is capable of detection times as short as five milliseconds. In the past, infrared detectors have been unsuitable for general applications because of the large number of false alarm sources found in the work place. However, when properly applied in controlled surroundings, they can provide reliable and effective protection.

A typical high speed IR detector consists of a cadmium selenide sensing element that is contained in a stainless steel housing. See Figure 5. By using a narrow bandpass infrared filter that is designed to minimize extraneous and ambient light sources, response is confined to the 0.75 to 0.85 micron range. This is the range that provides the fastest detector response. Figure 6 illustrates

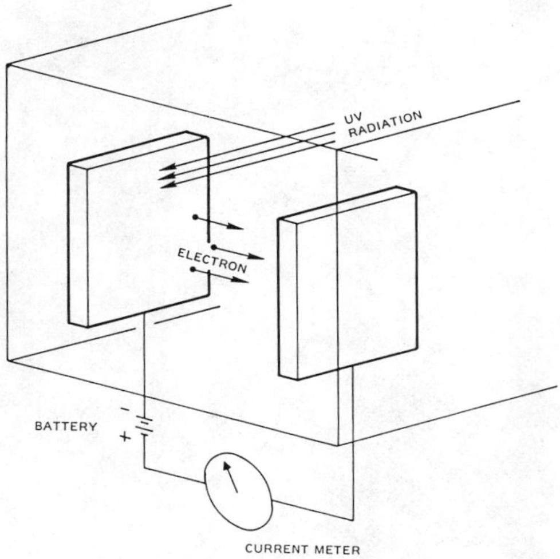

Figure 3. Ultraviolet Detector

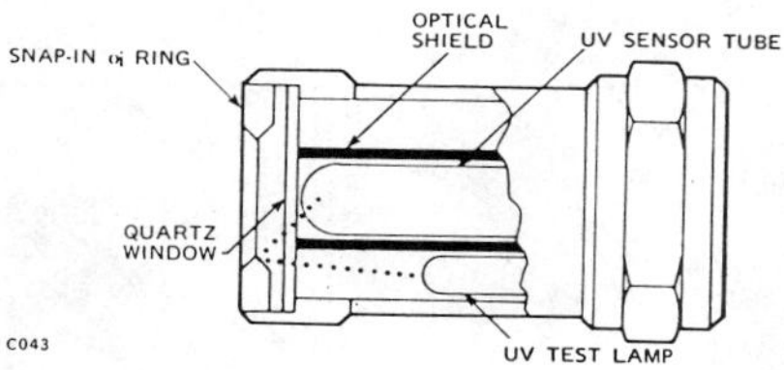

Figure 4. Optical Integrity

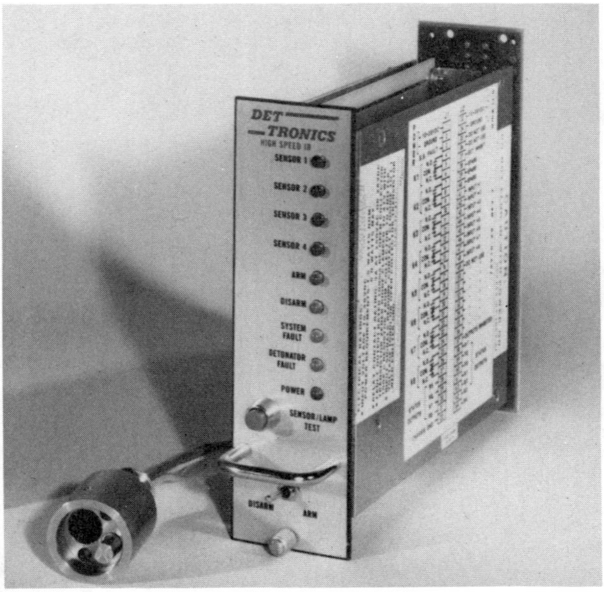

Figure 5. High Speed IR Detector and Controller

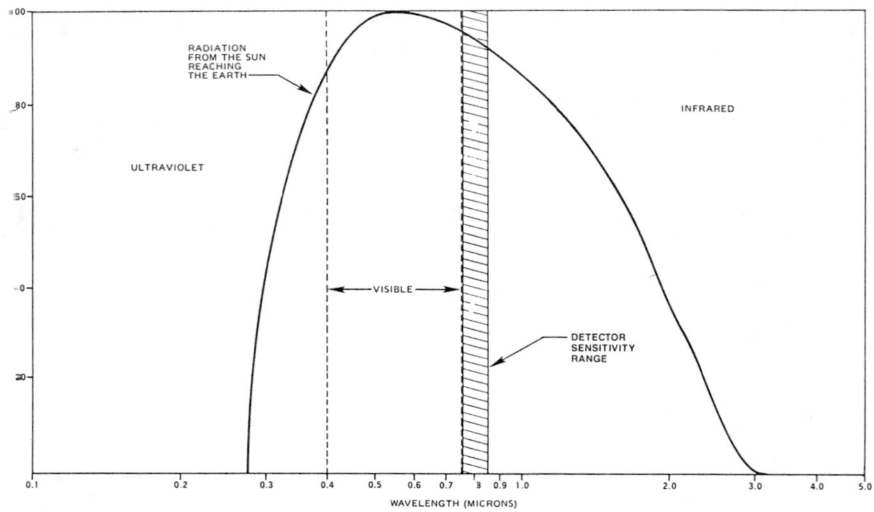

Figure 6. Response Range of Typical IR Detector

the general relationship between solar radiation at the surface of the earth and the spectral response range of a typical high speed IR detector. Note that this spectral response range includes false alarm sources such as the sun and artificial light. The detector also contains an infrared source within the enclosure. When a test signal is applied to the source, IR radiation is generated for testing the detector. Like the UV detector test, this test also checks the sensor and its wiring without resulting in a false alarm.

The IR detectors are usually connected to a controller that supplies power to the detectors and acts as a signal processor and output device. A typical controller monitors up to four detectors and energizes an output when any one of the detectors senses IR radiation that exceeds the alarm threshold level. The controller also contains the circuitry that checks the detectors and electrically supervises the interconnecting wiring to the explosive squibs or solenoid valves by trickling a small current through the external circuits.

Advantages. Like ultraviolet detectors, infrared detectors have their advantages and limitations. Several advantages of IR units make them valuable in certain installations:

1. They do not respond to the strong ultraviolet radiation from electric arc welding and lightning.
2. X-ray and gamma radiation do not extend to the infrared region, and single band IR units are not affected by them.
3. Smoke and/or vapors do not absorb radiation as significantly in the IR spectrum as in the UV spectrum. This makes devices of this type particularly useful when heavy smoke concentrations may accompany a fire. However, care must be taken that thick IR absorbing dusts are not part of the hazard.
4. An IR detector can "see" through substantially more contamination on its viewing window than a UV detector.
5. They are able to see hot ember-like fires typical of oxygen depleted areas.

Limitations. It is important to remember that the signal processing techniques necessary for reliable and stable detector operation may slow down the response time. In contrast, the requirements of the munitions industry have become more critical, requiring faster overall response times. The IR spectrum is broad and there are many sources of IR that radiate over the entire IR band. Typical are hot manifolds, boilers, processing vessels, engines and the sun itself. With some types of IR detectors the background radiation from a heat source can actually mask the presence of a fire and result in failure to respond. Attempts to use the well known flicker principle cannot be relied on to discriminate flame from background because of the amount of time needed for signal processing. To achieve the fast detection times needed, the IR detector cannot afford the luxury of the signal processing required to differentiate between the radiation emitted by fire and that emitted by blackbody radiation and ambient light. Therefore, high speed infrared sensors must be carefully isolated from possible false alarm sources. Such sources include the sun and other blackbody radiation sources, high intensity lights, flashbulbs, fluorescent and normal incandescent lighting.

Applications. The munitions industry has several applications suited for infrared detectors. Conveyor belts passing through large covered ducts and explosive and propellant mixers are examples of the controlled environment necessary for proper application.

Typical applications for these high speed IR detectors are characterized by strictly controlled, dark environments where a flash fire could originate. While simple high speed infrared systems have been available for several years, modern sensor and filter developments, coupled with state-of-the-art electronics, have resulted in systems tailored for the munitions industry. These systems are more selective within the electromagnetic spectrum, fast in response, and extremely flexible in application to suppression systems.

Typically, these systems are recommended to be used in combination with the appropriate ultraviolet systems, combining the advantages of ultraviolet for space protection with infrared for enclosed areas, as illustrated in Figure 7.

Response time of such systems is a function of ignition size, type of material, ambient air, fumes or vapor composition, distance and orientation of the fire source. When discussing the response times for detectors, it must be recognized that a far more important measurement is the speed of response for the entire detection and suppression system. For example, a high speed UV detector can detect a rapidly developing fire in approximately ten milliseconds under ideal conditions. In addition, however, the water extinguishing agent can require one hundred milliseconds or more to travel through the piping to the nozzle, and from the nozzle through the air to the fire. Thus it is important to realize that the speed of response of the detector is a small part of the total response time of the system.

Detonator Module

The Detonator Module is a control unit that is used with the UV and/or IR detection system to activate the water deluge system. When dealing with an entire fire detection system that utilizes more than one type of detector, a Detonator Module greatly expands the flexibility and capability of the system. An individual Detonator Module can accept multiple inputs from UV and IR controllers, other Detonator Modules, manual alarm stations, heat sensors, smoke detectors or any contact closure device. In the event of a fire, any of these devices will cause the internal fire circuitry of the module to activate the detonator circuit, sound alarms, and identify the zone that detected the fire. When properly used, a Detonator Module will add only one millisecond to the total system response time. See Figure 8 for an illustration of a fire detection system with a Detonator Module.

Reliable operation of the system is ensured by the ability of the Detonator Module to continuously monitor the input circuits and the detonator output circuits, to supervise the coil and wiring of the solenoid valve or squib, as well as to perform a self-test procedure to allow verification of other critical circuits.

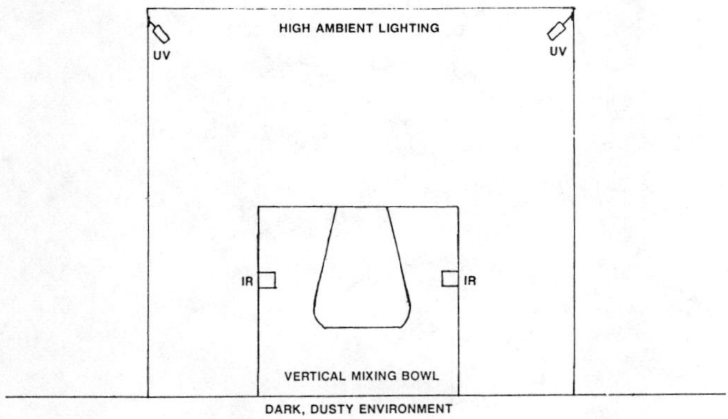

Figure 7. Typical Application Characteristics of UV and IR Detectors

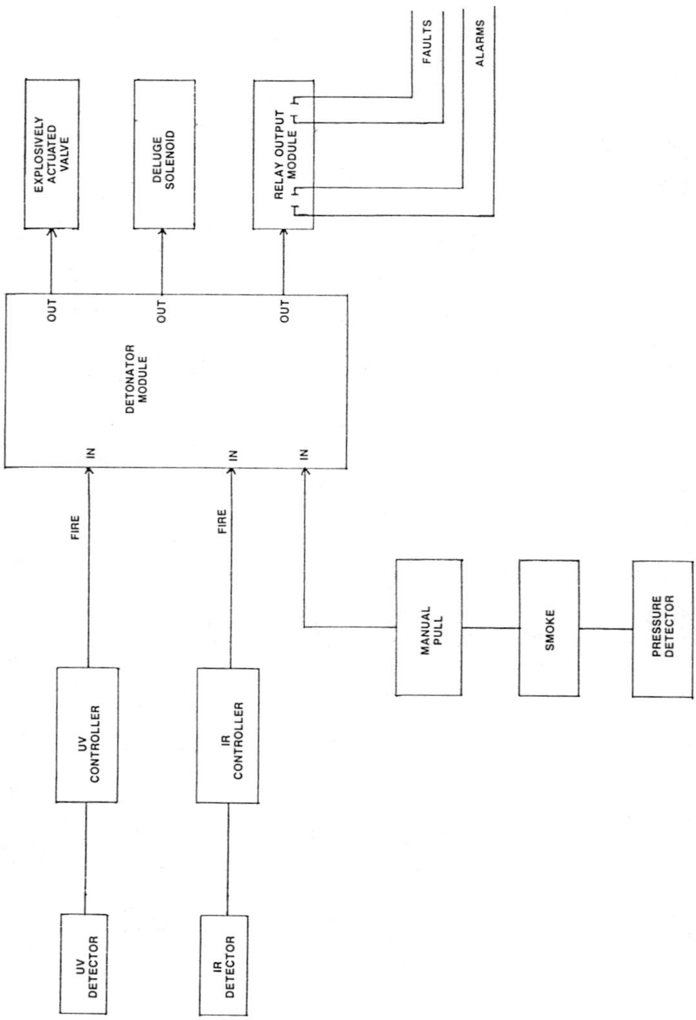

Figure 8. Fire Detection System with Detonator Module

Additional modules are available to output local or remote alarms as well as system fault conditons and are typically used to alert personnel to a fire condition or indicate the need for maintenance.

Extinguishing Agent

Over the years suppression tests indicate that water is the most successful and practical extinguishing agent for munitions type fires, since it cools to a point that prevents feed back of sufficient heat energy to maintain combustion. While it is important to get the water to the actual burning surface, it is not enough to wet a part of the surface only. The fire will burrow into the propellant and continue to burn, being shielded from water by the outer layer of water soaked material. This makes it necessary to apply the water rapidly, before the burrowing can occur. Another factor that makes rapid operation essential is that the water must reach the burning surface before the pressure of the combustion gases is sufficiently high to prevent the water from reaching the source of the fire. This requires that the system operate in a matter of milliseconds.

To successfully control a deflagration, large volumes of water must be applied quickly in a manner that will completely envelop the fuel. This is achieved by using a deluge system, by which water is simultaneously discharged from all outlets in the system, totally enveloping the hazard.

A typical high speed deluge system uses an electrically actuated (solenoid or squib) valve to initiate the flow of water from the nozzles. See Figure 9. The valve is positioned as close to the nozzles as possible. The piping between the valve and nozzles is fully primed and contains few if any air bubbles.

Figure 10 illustrates an explosively actuated valve. When the valve is in the set position, a plunger blocks the flow of water and is held in position by a shear pin and latch. Upon detection of a fire, a signal from the control panel fires the dual primers, causing the latch to swing to the tripped position. This breaks the shear pin, allowing the supply pressure to lift the plunger to the open position. With line pressure applied to the priming water, the caps or discs are blown off and water flows through the nozzles.

Factors Affecting System Operation Time

System response time can be divided into two phases. The first is the detection time, that is the time from the actual detection of the fire to the time that the signal is amplified and fires the primer in the water control valve or opens the solenoid valve. The second phase is the time required from primer firing or valve opening to the time water exits from the fire protection nozzles. The detection time is the fastest phase and under ideal conditions can be accomplished in as little as 10 milliseconds. The second phase, water delivery time, is the source of most of the time consumption.

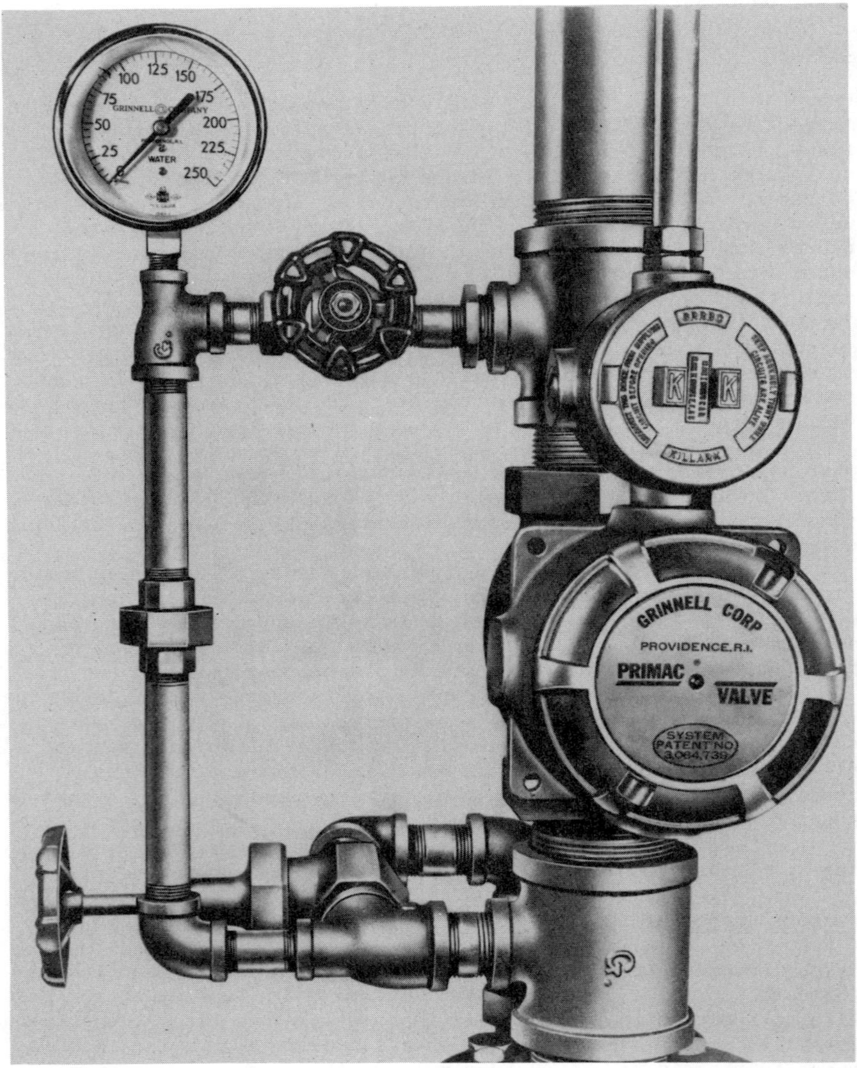

Figure 9. Explosively Actuated Valve

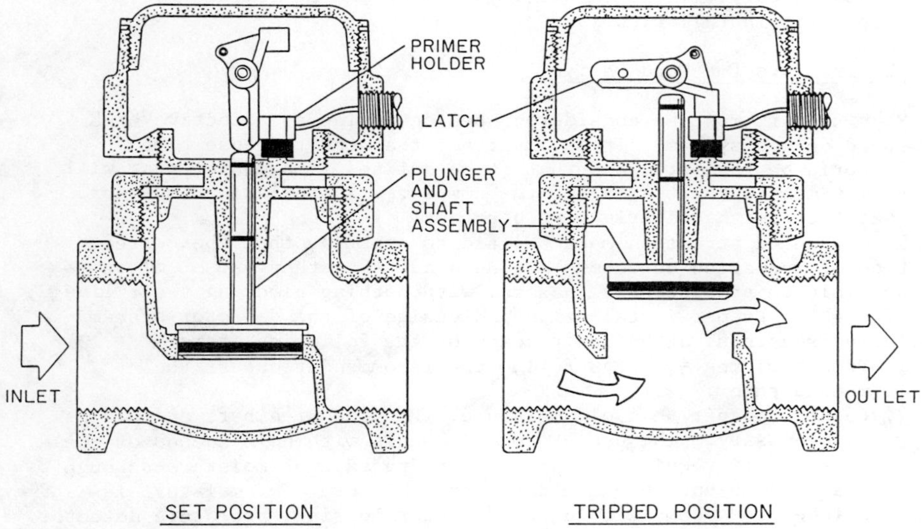

Figure 10. Water Discharge Valve

In order to optimize the system design, careful attention to the following is essential:

- Proper design and installation of radiation detectors.
- Adequate water supply pressure.
- Short and straight routes for the fire protection piping from water supply to nozzles.
- Length of pipe between control valves and nozzles as short as possible.
- Little or no air entrapped within the piping.
- Each installation is carefully designed and customized for the specific installation.

Requirements for Good Design

Many factors must be considered when designing an effective high speed deluge system. It is important that the response time criteria be realistic and that it be defined in a manner that will permit meaningful testing of the completed installation to ensure that the design criteria have been met.

Careful attention must be paid to ensuring that the correct type of radiation detectors are used and that they are as close as possible to the potential hazard, with nothing blocking their line of sight. Proper installation and design of the detector system includes careful attention to each of the following items:

1. Proper wiring - always follow the recommendations of the manufacturer.
2. Locate conduit to avoid moisture. Provide breathers and drains if necessary. The use of conduit seals within 18 inches of the detector is required to prevent the passage of moisture through the conduit and into the detector enclosure. If moisture is allowed to accumulate in the detector housing, premature detector failure can occur.
3. Is standby power needed?
4. Keep wiring runs as short as possible.
5. Consider the affects of lightning, welding, RFI, etc.
6. Locate the detectors as close as possible to the anticipated source of fire or explosion to increase signal strength and speed of response.
7. For best system performance and reliability, always use redundant detector coverage.

The water supply requirements must be determined. This involves estimating the maximum flow rate and the pressure required for adequate performance. The existing water supply and piping system should then be evaluated to determine whether or not it is able to meet these requirements. Remember that the water must be available instantaneously and that this cannot be accomplished by starting a fire pump.

Careful attention must also be paid to ensuring that no air bubbles are in the water piping, that the fastest possible water valves are utilized, and that the water nozzles are also as close as possible to the potential hazard to minimize the travel time of the water. These considerations will improve speed of response of the fire detection system to a much greater degree than improving the speed of the detector alone.

To design a successful high speed fire protection system, an engineer must take many factors into consideration. In addition, each individual installation typically has characteristics that require special attention. Because of the nature of the hazard involved and the need for such extremely fast response, the design of the high speed detection system is best left to a skilled expert. For only if the detection system and water deluge system work together to perform their functions in the shortest length of time will consistently reliable protection be possible.

RECEIVED March 6, 1987

Chapter 12

Ultra-High-Speed Fire Suppression for Explosives Facilities

Gary A. Fadorsen

Automatic Sprinkler Corporation of America, 1000 Edgerton Road, Broadview Heights, OH 44147

Advancements in Electronic Fire Detection in the past fifteen to twenty (15 to 20) years has made Ultra High Speed Deluge Systems for explosive facilities quite feasible and reliable. Since Detection has been covered in previous chapters, this chapter will focus mainly on Ultra High Speed Deluge Fire Suppression. Discussed are the three (3) most popular ultra high speed fire suppression systems presently used in explosive facilities. For the purpose of this paper, ultra high speed is defined as : A reaction time of less than 500 milliseconds, measured from the instant of fire detection to water flow at nozzle.

The evolution of Deluge Systems has been one of marked improvement. One of the first high speed deluge systems was the open head configuration that usually incorporated heat actuated detection, reaction time of this type system was approximately fifteen (15) seconds to two (2) minutes, depending on configuration and detection. Following this was the primed deluge system using optical flame detection (ultra-violet or infra-red). Reaction time of this type system could be as fast as one to two (1 to 2) seconds

During the "60's" the Squib actuated pre-primed deluge was developed. At least two major companies were supplying deluge systems in this configuration. The Squib actuated pre-primed system coupled with flame detection could respond well within the five hundred (500) millisecond range, thus providing the first ultra high speed deluge. This piping configuration consisted of single squib actuated deluge valve, primed piping and nozzles utilizing either caps or gold rupture discs to hold priming water. Common trade names for these systems are Primac and Spectronic. (A typical value is shown in Figure 1, typical piping layout is shown in Figure 2).

0097-6156/87/0345-0200$06.00/0
© 1987 American Chemical Society

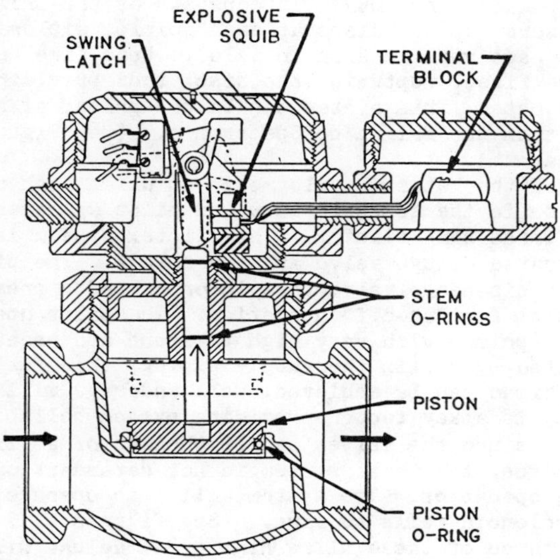

Figure 1. Primac Valve Cutaway

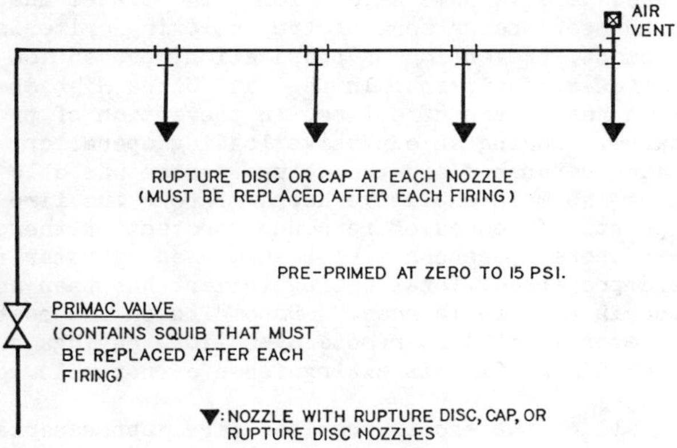

Figure 2. Typical Primac Configuration

Another system that falls into the ultra high speed category is also a squib actuated type, but the principle of operation varies greatly from the previous. It consists of pre-primed piping with high pressure rupture discs at each nozzle, explosive squibs are placed at each rupture disc (nozzle). Upon fire detection the squibs are fired, rupturing the disc, thus providing water at a very fast rate. This system can be pre-primed at a much higher pressure than the Primac or Spectronic. (See Figure 3 for typical layout).

One of the latest developments in ultra high speed suppression is the solenoid actuated pilot operated pre-primed deluge. Trade name Pilotex. The Pilotex system is essentially a pilot operated deluge valve at each nozzle. The pilot operated valve is a discharge valve that incorporated a pressure differential for "on-off" operation. The pilot operated deluge can be pre-primed with very high pressure and reaction time is not affected by air in the supply piping, thus fast and constant response times can be achieved, well under 50 milliseconds. Redundancy is a key factor providing system reliability and integrity, since the valves can be thought of as individual deluge valves, the total system is not dependent on one deluge valve for operation. The system will also operate even if all but one solenoid fails to fire. (See figures 4, 5, 6).

All three of these ultra high speed deluge will be discussed and compared in greater detail later in the text.

The justification for Ultra High Speed Deluge in an explosive facility would seem obvious,but there has been some debate on if and why this type of system is really necessary; the subject does deserve some discussion. The system must be designed to meet one or more of the following criteria: Total extinguishment, prevention of propagation, prevention of injury or protection of equipment. In the past Ultra High Speed Suppression has proven effective, in prevention of propagation for instance. During an explosive loading operation, the object being loaded detonated and the deluge system was able to prevent ignition of the main explosive hopper before the fire propagated to that point. In cases of personnel protection there have been many cases where operators have been doused by water and serious burns were prevented. Total extinguishment has been accomplished many times in past incidences. Depending on the cost of the equipment,even if it is a remote operation, savings can be substantial if the fire is extinguished or not allowed to propagate.

With all of the exotic chemical fire suppressants available today, one might wonder why water is used for high energy chemical mixtures, explosives, pyrotechnics, etc. Most all explosives, propellants, and pyrotechnic mixes contain the necessary oxygen for the burning process. Most high energy mixtures are a combination of a fuel and an oxidizer. The oxidizers provide the oxygen required for burning. Some examples of oxidizers are the nitrate and chlorate families, i.e., potassium nitrate, potassium perchlorate, barium nitrate, potassium chlorate, ammonium nitrate, etc. Because of these oxygen yielding substances, it is impossible to stop the

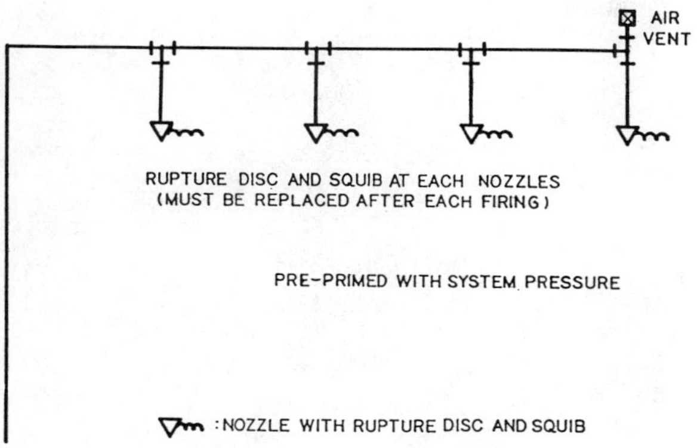

Figure 3. Typical Squib Actuated Rupture Disc Configuration

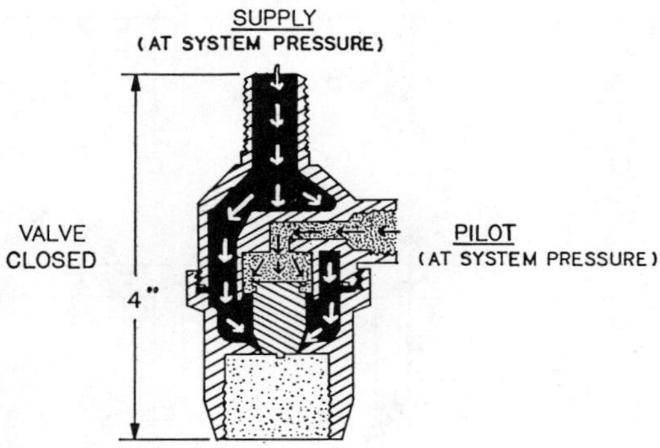

Figure 4. Pilotex Valve Cutaway (Closed)

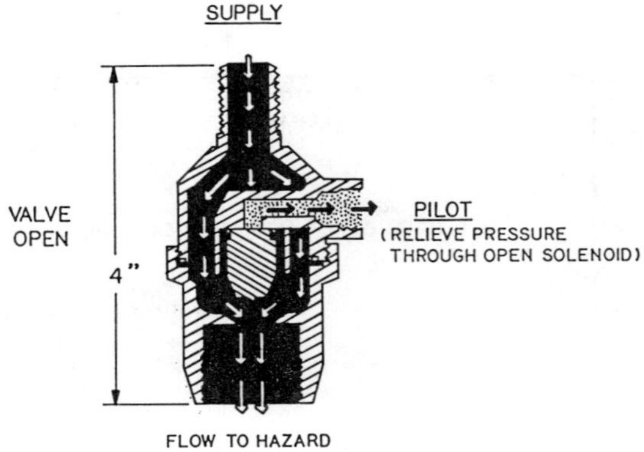

Figure 5. Pilotex Valve Cutaway (Open)

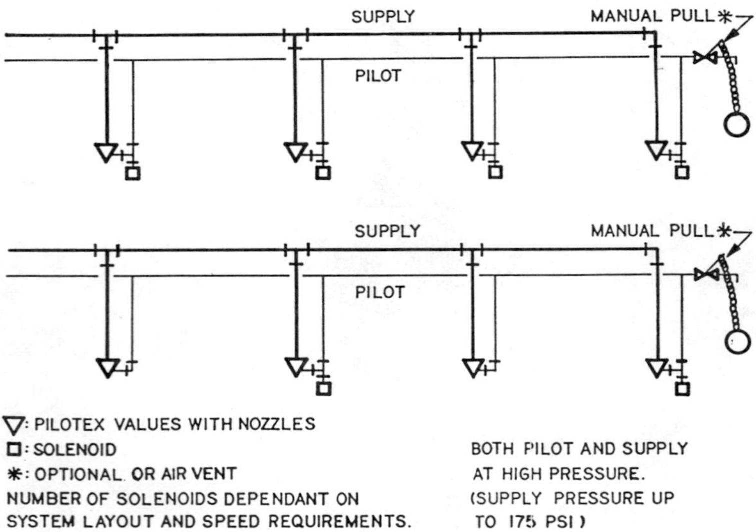

Figure 6. Typical Pilotex Configuration

propellant fire by suppressing the oxygen supply. Why water? It is generally agreed that cooling is a principal factor because it prevents feedback of sufficient heat energy to maintain combustion. It is of course desirable to get the water to the actual burning surface. However, it is not enough to wet the part of the surface, as the fire will burrow into the mixture and continue to burn, being shielded from the water by an outer layer of water soaked material. This makes it highly desirable to be able to apply the water rapidly before burrowing can occur. Another factor which makes rapid operation essential is that water must reach the burning surface before the pressure of combustion gases is sufficiently high to prevent water from reaching the source of the fire. This requires that the system operate in a mater of milliseconds. In some cases, especially with large bulk quantities of explosives, it my be necessary to flood the container from the bottom and the top or add a wetting agent to the water in the deluge system to allow penetration to the explosive. To summarize, the basic purpose of the water is to cool down and disperse the explosives or propellant. Applications for ultra-high speed suppression is as many and as varied as there are high energy products. Deluge systems have been used in primary high explosives such as mercury fulminate, lead azide, and DDNP. Secondary high explosives such as TNT, Tetryl, RDX, nitroglycerin, blasting gelatin and C4. Black powder is another very common application. Note that Sprinkler contractors should be notified not to use copper or brass fittings or components when protecting lead azide, due to the fact that lead azide in the presence of copper and moisture can become extremely sensitive copper azide. Ultra High Speed suppression is also well suited for the pyrotechnics and fireworks field. For example, magnesium teflon flares, colored stars, and smoke generating devices. In the case of magnesium teflon flares, the system could be used also to propel the burning flare away from the person to prevent burns.

The operations in an explosive facility also vary greatly and the system should be customized and geared towards the operation. The types of operations commonly seen in explosive facilities are weighing, pressing, pelletizing, propellant loading, melting, extrusion, mixing, blending, screening, sawing, granulating, drying, pouring, and machining. Each presents its own specific hazard and attention should be given to areas of ignition such as pinch-points, friction points, and areas where there is an operator working. The fire detectors and nozzles should be put as close to the hazard as possible. In many cases use dedicated nozzles to key-on specific problem areas, such as mixing bins, machining processes, extruder dies, etc. As mentioned before, determine what is required of your system. Is it to stop propagation, protect personnel, protect machinery? With this in mind, one can design a system that will effectively meet the needs. Each operation requires special consideration. For instance, during weighing, often with dry material, transfer and pouring of material can create dust and ignition, this is a good application for ultra-violet detection and high speed water spray. In the case of pressing and pelletizing, usually the goal here would be to prevent propagation. During the pressing and pelletizing operation, there is a good chance of explosion or ignition and the best bet would be to stop propagation to the bulk propellant. With propellant loading the transition of material from one vessel to another is a potential hazard. It is a good idea to have ultra-violet detection here and directed water spray. In the process of melting, usually these are closed melt kettles using steam for heat and often in this case infra-red detection with high speed nozzles directed into the kettle is a common configuration. With extrusion, the most likely point of ignition is where the material leaves the die. Again, keying the nozzles and detection at this point would help stop propagation. Mixing and blending are usually done in one of two (2) ways; within an open type mix or blending machine or a closed mixer. Depending on the type of machine, infra-red and/or U/V detection would be utilized, pressure detection is another option. The nozzles would be positioned accordingly. With screening, sawing and granulating, there is a good possibility for dust and sparks, key on the action.

System response time is a controversial issue that is often discussed but seldom settled. Probably the best and only concise way to determine if the deluge system is adequate is to run an actual fire test with the explosive or high energy material utilizing proposed detection and suppression system. Often this is not feasible for obvious reasons. The second most accurate method of time testing would be using high speed video cameras. Commonly these cameras record approximately one frame every eight (8) milliseconds, so what one does is record the event, play it back, count the frames and establish the response time. The advantage of this system is that you are able to see the propagation of the flame to the point of detection, the start of flow at the nozzle, and water spray as it progresses to the hazard, spray patterns can also be observed. This system is fine for a laboratory type evaluation but usually is not feasible for "in-field" application. Reasons being, the equipment runs from Fifty to Eighty thousand dollars ($50,000.00 to $80,000.00). Also, it is very bulky, often lighting is not adequate within the areas, the expense of providing the technicians and shipping the equipment is often prohibitive. So far, the most economical and reliable system for "in-field" time testing is a digital timer. Reaction time being defined as: beginning at instant of detection and stopping at flow from nozzle. The timer is started by a signal from detection control and is stopped by a flow switch connected at the nozzle. This seems to be acceptable by most authorities for testing deluge systems "in-field" and also for periodic maintenance testing.

Table I is a brief overview of available fast action deluge. The Primac is a squib actuated deluge valve. The system uses one large valve connected to a pre-primed piping system utilizing nozzles with end caps or rupture discs. In Primac Systems using rupture discs at the nozzle, the rupture discs are burst by water pressure not an explosive charge. The body of the Primac valve is that of a standard "globe" valve. The water seal is achieved by a piston entering the throat of the valve body. An "O" ring inserted in the same manner as a piston ring makes the piston watertight. The stem attached to the piston extends through the top of the valve. A swinging latch connecting this stem holds the valve in a closed position. The yoke supporting the latch is designed to accommodate a primer so positioned that when the primer detonates, the latch is forced off the stem and the water pressure under the piston opens the valve. NOTE: Be sure to keep stem "O" rings in good condition; a leak at this point may cause submersion of squib.

The explosive rupture disc system incorporates the same principle as Halon type explosive disc system, except that water is used as the extinguishing agent. In ultra high speed applications, there is a squib and rupture disc at each nozzle.

The Pilotex solenoid operated system does not use explosive squibs. It's principal of operation varies greatly from the previous two. When pilot pressure is relieved, all Pilotex valves connected to the one pilot light opens instantaneously and simultaneously. When the pilot pressure is restored, the nozzles close. A Pilotex valve consists of a two piece body threaded together and sealed with an "O" ring. The upper body has a half (1/2) inch NPT male connection for installation and standard pipeline fittings and a quarter (1/4) inch NPT female connection from the pilot line. It is through this pilot line connection that the cylinder and the poppet, that make up the differential valve, receive pilot pressure. The poppet has a teflon face which seats against the orifice located in the lower body half of the valve. The lower body is interchangeable to accommodate various types of discharge devices. Male adapters are often used where there is a need for flange mount or to directly flood a melt kettle or mixer. The female adapter is most often used with the Autospray nozzles. When the Pilotex valve is in its normally closed position, the poppet is held against the discharge orifice by the pressure within the poppet cylinder. When the pilot pressure drops, the main fire pressure overcomes the differential and forces the poppet up and instantly starts full discharge. When pilot pressure is restored, the poppet reseats, even against fire main pressure. Speed of the Pilotex system is not dependent on system size. Well under Fifty (50) milliseconds operation is guaranteed on all Pilotex system where such speeds are required.

With the various system available for the suppression high energy chemical fires, there is, in most cases a configuration suitable for almost any explosives, pyrotechnic or munitions facility.

Table I. Comparison of Ultra-High-Speed Deluge Features

	PRIMAC SPECTRONIC (NOTE A)	EXPLOSIVE RUPTURE DISC	PILOTEX	
COMPLETE ELECTRICAL SUPERVISION			X	(NOTE B)
RESPONSE TIME NOT AFFECTED BY AIR IN SUPPLY PIPE		X	X	
EXTRA PIPING FOR PILOT NOT NEEDED	X	X		
WIRING TO EACH SQUIB/SOLENOID NOT NEEDED	X			
NO RE-OCCURING COST OR REPLACEMENT PARTS NEEDED FOR RESET AFTER EACH FIRING			X	(NOTE C)
AUTOMATIC RESET FEATURE AVAILABLE			X	
SYSTEM DOES NOT REQUIRE EXPLOSIVES FOR OPERATION			X	
SYSTEM CAN BE SUPERVISED FOR HIGH PRESSURE PRIME		X	X	
SYSTEM CAN BE RESET AND BACK ON LINE IN LESS THAN 30 SECONDS			X	
INDEFINITE SHELF LIFE OF COMPONENTS			X	(NOTE D)
MECHANICAL MANUAL OPERATION AVAILABLE			X	(NOTE E)
ELECTRICAL PUSH-BUTTON RESET			X	(NOTE F)
EACH HEAD ACTS AS AN INDIVIDUAL DELUGE VALVE (SAFETY THRU REDUNDANCY)		X	X	
RESPONSE TIME NOT AFFECTED BY SYSTEM SIZE		X	X	
EXPLOSIVE SQUIBS NOT REQUIRED IN HAZARD AREA	X		X	
COMPATIBLE WITH ALL FORMS OF DETECTION	X	X	X	
SYSTEM CAN BE PRE-PRIMED WITH HIGH PRESSURE		X	X	

NOTE A: Due to innovations with Pilotex, Spectronic system was obsoleted by manufacturer.

NOTE B: Solenoids are supervised for short, opens and grounds. On a Squib operated system, the igniter wire can be supervised but condition of explosive is not known.

NOTE C: After firing of the squib operated systems disc or caps must be replaced/squibs must be replaced.

NOTE D: Squibs have a shelf life and should be periodically replaced.

NOTE E: Mechanical manual release is possible even in the event of total power failure (including loss of primary power and battery back-up).

NOTE F: Pushing reset button closes solenoid valves thus re-setting system.

Literature Cited

(1)
Pyrotechnics;, George W. Weingart
(2)
Chemistry of Pyrotechnics, Basic Principles and Theory;, John A. Conkling
(3)
Survey of Sensitivity Characteristics of Typical Delay Ignitor Flash and Signal-Type Pyrotechnic Compounds; Joseph Kristal and Seymour M. Kaye; Picatinny Arsenal, Dover New Jersey;439383
(4)
Fireworks from a Physical Standpoint;Dr.Takeo Shimizu, Kawagoe-Shi, Japan
(5)
The Chemistry of Powder and Explosives;, Tenney L Davis
(6)
The Safe Practices Manual for Manufacturing, Transportation, Storage and Use of Pyrotechnics;publication of The Department of Health,Education and Welfare,bulletin #PB-297807
(7)
Fire Extinguishing Devices in Propellant Plants;, Peter R. Weibel, Swiss Federal Propellant Plant, Wimmis, Switzerland
(8)
The Minutes of the Rapid Action Fire Protection Seminar, AMC,AMCCOM Safety Office; 23rd and 24th,October, 1984
(9)
Pilotex Ultra High Speed Deluge Fire Protection for Munitions Explosives Pyrotechnics ; Gary A. Fadorsen, "Automatic" Sprinkler Corporation of America
(10) The Loading, Assemblying and Packing of M42 Grenades;"Are We Doing It Right?";Donald R. Kennedy
11) Recent Developments in High Speed Optical Detection Systems for Pyrotechnic and Munitions Processing from the

International Pyrotechnic Society Seminar in Vail, Colorado; July 6 thru 11, 1986; Ken M. Klapmeier of Detector Electronics Corporation, Minneapolis, Minnesota
(12) Ultra High Speed Deluge System for the DDESP Seminar; Bob Lloyd,AMCCOM, Rock Island, Illinois
(13) False Alarm Reduction in Industrial Flame Detection; Roger A. Wendt, Armteck Industries, Manchester, New Hampshire
(14) Ultra-Violet Deluge Study XX-61; Day & Zimmerman
(15) Explosives and Demolitions,Department of the Army Field Manual,FM5-25
(16) Engineering Guide for Fire Protection and Detection Systems at Army Amuunition Plants, Vol. 1;Lewis Joblove, Manuel Avelar and Norval Dobbs, of Ammann and Whitney

RECEIVED March 6, 1987

CHEMICAL EFFECTS

Chapter 13

Systematic Approach for Safely Designing a Chemical Surety Materiel Laboratory

George E. Collins, Jr.

Chemical Research Development and Engineering Center, Attn: SMCCR-SFC, Aberdeen Proving Ground, MD 21010-5423

This article shows, through example, how established system safety concepts can be used to develop safety criteria for the design of a chemical surety materiel laboratory. This systematic approach, when applied as described in this article, results in a laboratory dedicated to achieve mission objectives in an environment relatively free of inherent hazards for the least number of dollars.

Facility System Safety (FSS), which is the application of system safety concepts to the facility acquisition process, has recently gained acceptance throughout the Department of Defense and most recently within the Department of Army with the conception of SAFEARMY 1990. The Army's goal is to: fully integrate the total system safety, human factors, and health hazard assessments into continuous comprehensive evaluation of selected systems and facilities. The Chemical Research Development and Engineering Center (CRDEC) has mandated appropriate levels of system safety throughout the lifecycle of facility development for many reasons. These include:

1. Optimum safety and health are required to prevent personal injury to chemical surety agents. Facility System Safety is one avenue used to achieve optimum safety and health in operations that deal with these agents.
2. FSS is a proactive approach which will reduce inconsistencies during the facility acquisition process. This results in a more mission responsive facility that is less expensive.

The intended purpose of this article is to demonstrate, through specific examples, how FSS can be applied to the design/construction/operation of a chemical surety materiel laboratory. The laboratory under study is a 32 million dollar Military Construction, Army (MCA) project designed to replace aging facilities which are currently utilized to perform daily Chemical Surety Materiel (CSM) operations. For the purpose of this article, CSM is defined as a

This chapter not subject to U.S. copyright
Published 1987 American Chemical Society

chemical compound used in military operations to kill, seriously injure or incapacitate a person through chemical properties. This article demonstrates the methods used in identifying, analyzing and ultimately eliminating or reducing the effect of a hazard on the facility, equipment and personnel.

Facility System Safety Overview. The process of applying system safety to the facility acquisition process can be divided into the following tasks:

1. Risk Categorization
2. Preliminary Hazard List
3. Preliminary Hazard Analysis
4. Design Considerations

The remainder of this article will involve a description of each of these tasks followed by an example of how the task was applied to the design of this CSM laboratory.

Risk Categorization. The first step in this process is to clearly define the risk associated with the operation of this laboratory. This step includes a brief description of the operation followed by a risk assessment and a recommendation on the level of system safety required.

Laboratory Description. The laboratory under consideration will conduct diversified chemical surety materiel laboratory operations. These materials are anticholinergic agents and are extremely lethal in small concentrations. The recommended permissible airborne exposure concentration for some of these agents is 0.0001 mg/m3 (2 x 10-5 ppm). Two personnel are required, as a minimum, to perform this operation.

Assessment. An analysis of the hazards present in this laboratory show the most significant hazard to be the release of vapor CSM from engineering controls and into the workplace. The significance of this hazard mandates further efforts in system safety in the form of a Preliminary Hazard List (PHL) and a Preliminary Hazard Analysis (PHA). The user must in this instance take an active role in the design review process.

Preliminary Hazard List. Once the risk categorization is completed, the next step is to develop a PHL. The PHL is a user generated listing of hazards which must be controlled. The user must, at this stage, assign a risk assessment code to each hazard and establish any further requirements for analyses (the methodology used in the development of risk assessment codes in this article is shown as Figure 1). As a minimum the user should use the following sources of information for PHL development:

1. Material Safety Data Sheets
2. Feasibility Studies
3. Project Development Brochures
4. Standing Operating Procedures
5. Operator Interviews

Hazard Severity

(1) Category I - Catastrophic: May cause death or loss of a facility.
(2) Category II - Critical: May cause severe injury, severe occupational illness, or major property damage.
(3) Category III - Marginal: May cause minor injury, minor occupational illness, or minor property damage.
(4) Category IV - Negligible: Probably would not affect personnel safety or health, but is nevertheless in violation of specific standards.

Mishap Probability

(1) Subcategory A - Likely to occur immediately.
(2) Subcategory B - Probably will occur in time.
(3) Subcateogry C - May occur in time.
(4) Subcategory D - Unlikely to occur.

Risk Assessment Code

		Mishap Probability			
		A	B	C	D
Hazard Severity	I	1	1	2	3
	II	1	2	3	4
	III	2	3	4	5
	IV	3	4	5	5

Figure 1. Risk Assessment

Preliminary Hazard List Description. The incorporation of this information into a PHL entry is shown as Table I. This entry describes; the nature of the hazardous event (column 1), why or how the hazard may result in a mishap (column 2), the effects on operating personnel, equipment, and the facility (column 3), the risk assessment code assigned to the uncontrolled hazard (column 4) and any comments the originator may have (column 5).

Preliminary Hazard Analysis. The next step in the process is the development of a PHA. This analysis is the core of the FSS program and as such is vital in eliminating or reducing the inherent hazards associated with this laboratory operation. The PHA is used to further analyze the data identified in the PHL. This enhances the hazard control data base and provides specific recommended corrective action for the resolution of hazardous conditions. A combination of the informational sources used in the PHL development and any additional design information should be used in PHA development.

Table I. Preliminary Hazard List

COLUMN 1	COLUMN 2	COLUMN 3	COLUMN 4	COLUMN 5
HAZARDOUS EVENTS	CAUSAL FACTORS	EFFECTS	RISK ASS. CODE	COMMENTS
Release of vapor CSM from lab hood and into workplace or atmosphere.	1. Power failure	1. Loss or lab hood capture. Release of CSM into workplace. Personnel injury or death. System/facility damage minimal.	I A 1	None
	2. Mech. exhaust fan failure	2. Same as #1 above.	I B 1	None
	3. Poor lab hood capture (Design)	3. Turbulence may result in small release of CSM into workplace. Personnel injury or death could result. System/facility damage minimal.	I B 1	None
	4. Operator error	4. Judgement errors could result in an inadvertent release of CSM into the workplace. Personnel injury or death could result. System/facility damage minimal.	I B 1	None
	5. Filters do not remove CSM from exhaust	5. Personnel injury to people surrounding the facility. System/facility damage minimal. Adverse publicity.	II C 3	Scenario less likely and severe due to dilution factor.
	6. Exhaust ductwork not properly sealed	6. Small concentrations CSM in the workplace possible in the event the exhaust system were to go positive. Personnel injury or death possible. System/facility damage minimal.	I C 2	Scenario less likely due to additional requirement for system to go positive.

Table II. Preliminary Hazard Analysis

COLUMN 6 RECOMMENDED ACTIONS	COLUMN 7 CONTROLLED RISK ASS. CODE	COLUMN 8 STANDARDS
<u>Causal Factor #1</u>: a.) Emergency generator system shall be installed to automatically initiate in the event of a power failure, system phasing shall be accomplished in a manner which will not permit the occurrence of a hazardous condition. b.) Laboratory hoods must be equipped with a mechanism to warn operators of emergency power status and hood function. c.) Standing Opeating Procedures should contain provisions for the curtailment of operations, immediate masking and evacuation from areas that experience power failures.	IV D 5	DOD 6055.9-STD AMCR 385-102 CRDECR 385-1
<u>Causal Factor #2</u>: a.) Two alternatives are available to prevent a hazardous condition from occurring in the event of a mechanical failure. These include: (1) Redundant exhaust fan units, (2) Procedural controls which require curtailment of operations, donning of protective masks and immediate evacuation during ventilation loss. b.) Laboratory hoods shall be equipped with a means to warn operators of improper ventilation system functioning.	IV D 5	DOD 6055.9-STD AMCR 385-102 CRDECR 385-1 LOCAL SOPs
<u>Causal Factor #3</u>: a.) Laboratory hoods must be located away from: - Main traffic aisles and doorways - Adjacent walls and operable windows - Cross drafts exceeding 30 lfpm - Heating Units - Exits. b.) Laboratory hoods must perform as follows: - Average inward face velocity of 100 lfpm +/- 10% with the velocity at any point not deviating from the average face velocity by more than 20%.	IV D 5	AMCR 385-102 AEHA Technical Guide #30 CRDECR 385-1

Table II. Continued

COLUMN 6	COLUMN 7	COLUMN 8
RECOMMENDED ACTIONS	CONTROLLED RISK ASS. CODE	STANDARDS
Causal Factor #3 (Continued): c.) Operators must be trained in proper operation within a laboratory hood.		
Causal Factor #4: a.) Operating personnel must be properly trained. b.) Operating personnel must wear appropriate protective clothing. c.) Operating personnel must work under a properly approved SOP.	IV D 5	CRDECR 385-1
Causal Factor #5: a.) Exhaust filtration system shall meet CSL SOP 70-18.	IV D 5	CSL SOP 70-18 CRDECR 385-1
Causal Factor #6: a.) Ductwork shall be sealed to preclude leakage. b.) All joints shall be seamless welded. c.) Ductwork shall be capable of withstanding 16 inches water column vacuum and 25 inches water column positive pressure.	IV D 5	DOD 6055.9-STD CRDECR 385-1

Table III. Hazard Tracking Log

COLUMN 9	COLUMN 10	COLUMN 11	COLUMN 12
ACTION TAKEN	TRANSFER	DESIGN CERTIFICATION	CONSTRUCTION CERTIFICATION
CAUSAL FACTOR #1:			
a.) Emergency generator installed and properly phased	Drawing #:099 Specification Section #09991	Mr. Smith	Mr. Jones
b.) Laboratory hoods equipped with warning devices to notify operator of power loss	Drawing #:061 Specification Section #08001	Mr. Smith	Mr. Jones
c.) Installation notified of finding	Disposition Form sent 6 Jan 86 to safety office	-----	-----
CAUSAL FACTOR #2:			
a.) Installation safety office determines need to go with procedural controls. SOPs will be developed accordingly.	Disposition Form 10 Jan 86	-----	-----
b.) Laboratories equipped with warning devices to notify operators of ventilation system failure	Drawing #:061 Specification Section #08001	Mr. Smith	Mr. Jones
CAUSAL FACTOR #3:			
a.) Lab hoods meet the following: Away from: - Main traffic aisles - Doorways and windows - Adjacent walls - Cross drafts > 30 lfpm - Heating units - Exits	Drawing #:045	Mr. Smith	Mr. Jones

Table III. Continued

COLUMN 9 ACTION TAKEN	COLUMN 10 TRANSFER	COLUMN 11 DESIGN CERTIFICATION	COLUMN 12 CONSTRUCTION CERTIFICATION
CAUSAL FACTOR #3: (Continued) b.) Lab hoods perform as follows: - Average face velocity 100 lfpm +/- 10%. No single reading deviating from average by 20% - Smoke testing did not result in a release of visible smoke	Drawing #:046 Specification Section #07010	Mr. Smith	Mr. Jones
c.) Installation notifed of requirement for proper training of operators	Disposition Form dated 25 Mar 86	-----	-----
CAUSAL FACTOR #4: Installation responsibility	Installation notified 25 Mar 86	-----	-----
CAUSAL FACTOR #5: Exhaust system complies with CSL SOP 70-18	Specification Section #01001	Mr. Smith	Mr. Jones
CAUSAL FACTOR #6: Ductwork properly sealed and tested	Specification Section #02000	Mr. Smith	Mr. Jones
	Disposition Form dated 25 Mar 86	-----	-----

Preliminary Hazard Analysis Description. The incorporation of this information into a PHA entry is shown as Table II. This entry describes; the proposed actions needed to eliminate or control the hazard (column 6), the risk assessment code assigned after controls (column 7), and the identification of applicable codes and standards (column 8).

Hazard Tracking Log. In addition to the above analysis, a hazard tracking log (HTL) should be maintained. This log is to ensure all open loops are closed and ensures the appropriate level of management is identified as being involved in the acceptance of risk. This log should be initiated during the design phase and maintained throughout construction. As this facility is not at the design stage at the time of publication, a simulated HTL was used and is shown at Table III. This entry describes: the specific action taken to eliminate, control or accept the hazard (column 9), the reference of the blueprint/drawing numbers or other documents that address the action taken (column 10), name of individual closing out the action on design (column 11), and the name of the individual closing out the action during construction (column 12). The information contained in this log is proposed because the laboratory is in the design stage of development.

Laboratory Design Considerations. As a result of this effort, detailed safety design considerations can be developed to preclude the release of lethal concentrations of vapor CSM into the workplace. This will minimize the potential for death or serious injury to our research scientists. A summary of these requirements is shown in Appendix A.

Conclusions. The effort put forth in FSS for this laboratory has many benefits. Most noteworthy are:

1. Safest possible laboratory
2. More mission responsive facility
3. Less expensive facility

This article is a step in the direction we must all head toward and that is total system safety for facilities to reduce inherent hazards associated with their operation.

Literature Cited

1. Chemical Systems Laboratory Standing Operating Procedure (CSL SOP) 70-18, 10 Nov 82, Exhaust Ventilating Systems.
2. Department of Defense (DOD) 6055.9 - STD, July 1984, Ammunition and Explosives Safety Standards.
3. Army Materiel Command Regulation (AMCR) 385-102, 6 May 1982, Safety Regulations for Chemical Agents GB and VX.
4. Chemical Research Development and Engineering Center Regulation (CRDECR) 385-1, 15 Aug 1986, Chemical and Occupational Safety and Health Program.

APPENDIX A

Laboratory Design Considerations for Protection Against Vapor Chemical Surety Materiel Exposure

A. Electrical Design Considerations (Causal Factor #1):

1. Emergency generator systems will be installed to service the following:

   - Exhaust ventilation fans
   - Make-up air handling units
   - Critical operating equipment
   - Emergency lighting
   - All emergency alarm systems

2. Diesel-powered generators will be used. The emergency generator will be sized to handle 100% of the connected emergency load.

3. Start-up of the exhaust ventilation system and critical equipment must be sequenced to prevent a hazardous condition. In addition, the starting of the supply air handling unit and the exhaust fan services each room shall initiate simultaneously to avoid placing the room under positive pressure. Automatic transfer switching will be used.

B. Warning Systems (Causal Factor #1 & 2):

1. Facility will be equipped with a master control panel and alarms which permits functional verification of the exhaust blowers, filters, make-up air supply systems, fire control systems and waste treatment processes.

2. Laboratory hoods will be equipped with audible and visual alarms which will be designed to initiate when the average inward face velocity falls below 90 linear feet per minute.

3. Visible alarms must be located so they can be readily seen by personnel while working at the exhaust hood.

4. A test switch must be installed on all alarms which will permit the operator to verify that the light has not burned out and the sound alarm will function. This test must be performed while ventilation system is in full operation.

C. Laboratory Hood Location (Causal Factor #3):

1. Laboratory hoods must be located away from:

   - Heavy traffic aisles
   - Doorways

- Adjacent walls
- Crossdrafts that exceed 30 lfpm
- Heating units

2. Sidewall registers and conventional ceiling diffusers shall not be used for laboratory air supply.

3. Perforated ceiling panels shall be used so that distribution of supply air is three feet minimum from the front face of the hood. The exit velocity from these panels shall not exceed 35 lfpm.

D. Laboratory Hood Performance (Causal Factor #3):

1. Laboratory hoods shall have an average inward face velocity of 100 lfpm +/- 10% with the velocity at any point not deviating from the average face velocity by more than 20%

2. Leakage testing must be done with 30 second or one minute smoke candles placed approximately 20 centimeters inside the hood. Any visible escape of smoke should be considered indicative of unacceptable performance.

3. Laboratory hoods shall be designed as deep and low in height as practical. Rough wall surfaces and recesses in walls and work surfaces are unacceptable.

4. The location of sash tracks and the number of baffles and slots provided are integral to the proper containment of materials.

5. Laboratory hoods will be equipped with a 20 centimeter line taken from the face of the hood. No CSM contaminated equipment should be placed in front of this line during operations.

E. Exhaust Ventilation/Filtration System (Causal Factor #5):

1. All laboratory exhaust air shall be exhausted through a filtration system which complies with CSL SOP 70-18. These systems have been proven to be effective in removing CSM vapor from an exiting airstream.

2. Ventilation exhaust shall not be recirculated.

3. Instrumentation shall be required to monitor and control the airflow through the filter system. Instrumentation shall provide a means to monitor overall pressure drop as well as the pressure drop between each filter element.

4. The filter system shall include a series redundant-parallel Chemical Biological Radiological (CBR) filter assembly with a capability of placing a detector between the adsorber banks to warn of "breakthrough". The system shall provide accessibility to filters for repairs, maintenance and leak testing.

5. The filter system shall be as follows:

Hood - Prefilter - HEPA - Adsorber - Adsorber - HEPA - Exhaust

6. Exhaust stacks shall be designed and constructed to ensure good dispersion of exhaust air to the atmosphere thereby preventing recirculation.

F. Exhaust Ductwork (Causal Factor #6):

1. All ductwork shall be round, and welded with flange connections.

2. Ductwork shall be designed to facilitate dismantling and to minimize the release of contamination to adjacent areas with bagging or other approved means.

RECEIVED March 6, 1987

Chapter 14

Laboratory Design

Frances H. Cohen

Oneil M. Banks, Bel Air, MD 21014

Research laboratories are very unique facilities which require a great deal of preparation and coordination to produce a proper design. Much like the research that will be performed in the facility, each laboratory has specific needs and requirements. The primary considerations in the design of a research laboratory include the ventilation system, types of research and associated equipment, and safety and health of the work environment. Each of these primary consideration are of equal importance to the development of a successful design.

A safe and healthful work environment is a crucial requirement of a research laboratory. This consideration is the most often overlooked, yet it is intertwined with all aspects of building design and operation. Protection of the buildings occupants includes not only fire safety aspects as defined in the National Fire Protection Association Life Safety Code, but in the breathing air quality. Therefore, the materials of design, means of egress, and ventilation system should be the first subjects considered during the design process.

Just as laboratories are unique from other buildings, so are their ventilation systems. The laboratory chemical fume hood is the primary engineering control used to protect workers from potential serious exposures to toxic substances, yet they are often the last furnishings considered.

The subsequent sections of this chapter will outline a team approach to laboratory design.

THE TEAM

In order to properly consider all aspects of laboratory design and keep each specialty in perspective a design team should be assembled. This team consists of individuals from each discipline involved in the design: engineering; research program; safety and health; and an architect/engineering (A/E) firm. Tradition is maintained at this point as the engineer becomes the focal point of the team. This individual is responsible for the total project coordination and selection of the A/E. Further, this individual must, sometimes, act

0097-6156/87/0345-0224$06.00/0
© 1987 American Chemical Society

as a mediator when disagreements arise. Representation is limited to a few scientists. They, in turn, usually develop their own research program subcommittee. The extent of this subcommittee is dependent of the size of the building and the number of different research programs that will occupy the building.

Health and safety considerations are addressed jointly by an industrial hygienist and a safety specialist. These are the individuals that are "the authority having jurisdiction" as referenced by the National Fire Protection Association. Because of the unique nature of many research laboratories, it is not always possible to adhere strictly to the NFPA Codes and these individual must use their professional judgement in applying the intent of the Codes.

Once the team is assembled, it is important to have a "kick-off" or pre-design meeting so that each representative is given the opportunity to present their needs and requirements. The remainder of this chapter will be devoted to health and safety requirements in the design of a research laboratory.

DEFINITION

There is probably nothing more confusing than the definition of a laboratory. For the sake of consistency in this chapter a laboratory is defined as a building, space, room, equipment, or operation used for testing, analysis, research, instruction, or similar activities. To further explain this definition, a room is considered a laboratory if any of the following exist:

1. fume hood/biosafety cabinet
2. gas cylinders
3. use or storage of chemicals with any of the following properties;
    a. flammable
    b. combustible
    c. explosive
    d. water sensitive
    e. caustic
    f. corrosive
    g. high or unknown toxicity
    h. carcinogen/mutagen/teratogen
4. biohazardous material
5. grinding operations
6. radioisotopes/radioactive sources

CODES AND SPECIAL REQUIREMENTS

At the outset of involvement in laboratory design it is incumbent for the health and safety specialists to designate those codes, regulations, and special requirements they consider essential to produce a safe and healthful work environment. All to often the A/E will choose a standard building code to follow. These codes, while appropriate for office buildings, do not address the necessary life safety requirements necessary for laboratories.

Typically, the codes and regulations required for proper health and safety in laboratory design are the National Fire Protection

Association Code, Occupational Safety and Health Administration Standards, Environmental Protection Agency Regulations, National Institutes of Health/Centers for Disease Control Biosafety Guidelines, Nuclear Regulatory Commission Regulations, American National Standards Institute Standards, American Conference of Governmental Industrial Hygienists Manual on Industrial Ventilation, and specific written policy of the agency/company planning the design. It is becoming more commonplace for individual companies to develop specific requirements for ventilation systems, biosafety facilities, radiological safety, animal facilities, and performance standards for equipment.

Under the NFPA Life Safety Code each building is given an occupancy classification. Laboratory structures are usually classified as "industrial" with "high hazard contents." When a building is designed for mixed occupancies such as offices and clinical areas, separate classifications con be assigned if separate safeguards are provided in accordance with the Code. The designation of a buildings occupancy classification is important to the selection of building materials, placement of the mechanical room, and egress design, location and number.

VENTILATION SYSTEMS

There are two main types of ventilation systems; constant volume and variable volume. Both systems can be either 100% fresh air or recirculating. The type of system that is selected should be carefully chosen with safety and health as the primary consideration. Constant volume systems deliver a preset volume of air over a specified temperature and humidity range. Variable volume systems deliver variable amounts of air which are determined by temperature change, air needed (i.e., use of fume hoods), and by pressure differentials. Constant volume systems are dependable and require little maintenance, but are not energy efficient. Variable volume systems are usually energy efficient, but require sophisticated technology and scheduled preventive maintenance. Only recently has the technology been developed to properly implement variable volume systems. There are numerous pros and cons for selection of either system which will not be discussed at this time. However, the team must consider funds available for the project, maintenance provisions, and current and future research needs before making a selection.

In a mixed occupancy building it is wise to consider the design of separate ventilation systems for laboratory areas, areas servicing the public, animal holding areas, and administrative offices. Although this approach adds additional cost to both the design and construction of the building, it allows for selection of different systems in each area and increases the flexibility of the research functions. Also, the use of separate ventilation systems allows for the use of more energy efficient systems in those areas where air recirculation can be employed safely. For example, if the ventilation system for the laboratory area is properly designed, the addition of another fume hood can be achieved without redesign or any effect to other areas of the building. Animal holding areas create their own unique requirements depending on the species and type of research to be performed. A separate ventilation system allows the flexibility to change for these requirements as they arise without effecting other areas of the building. Requirements for animal holding areas will be discussed further under Animal Facilities. Further, the use of variable air volume recirculating systems in offices is an effective way to save energy and provide a healthful work environment if, using a moderate flow rate, 20% fresh air is introduced into the system.

Design of laboratory ventilation systems should be approached with the realization that the laboratory can be the single most dangerous workplace. Strict adherence to NFPA 45 - Fire Protection for Laboratories Using Chemicals is advised. Recirculation of laboratory air should be prohibited as it poses both a fire safety problem and a potential health hazard. Recirculating systems allow for more rapid spread of fire to other areas. More importantly, in the event of a toxic chemical spill, recirculating systems spread the contamination throughout the laboratory and do not provide the necessary exhausting capacity to remove the chemical from the environment that a 100% fresh air system will perform.

Although room changes of air per hour is not a very technical means of determining that enough air is supplied to a laboratory area, it is a term which is easily understandable. For most laboratory applications, eight to twelve room changes per hour are adequate to provide proper dilution ventilation. Laboratories designed for biocontainment require a minimum of ten room changes of air per hour.

For both fire safety, health considerations, and proper functioning of fume hoods the air pressure of laboratory areas must be negative relative to surrounding areas. The only exception to this is for certain biocontainment applications. These applications usually require very specific ventilation requirements which will not be addressed. Also, all laboratory ventilation systems, especially fume hoods, should incorporate low flow warning devices.

As stated earlier, the laboratory chemical fume hood is the single most important engineering control in the laboratory for the protection of workers from exposures to toxic substances. While this statement usually receives widespread approval, the lack of attention paid to fume hood design specifications and location within the laboratory is truly amazing. While a fume hood is a very substantial piece of equipment it's proper functioning is dependent on delicate placement and balancing. Recent developments in research on fume hood face velocities has shown that face velocities as low as 75 feet per minute (fpm) are sufficient for the handling of volatile materials. With this reduction of face velocities it becomes more important than ever to place fume hood away from traffic areas and supply air diffusers. When walking, the average person creates turbulence of approximately 250-300 fpm. Slight movement, such as breathing can create turbulence as high as 25 fpm. Air supply diffusers generally supply air at 100 fpm or higher. Therefore, it is easy to see how these otherwise insignificant events can totally disrupt the proper operation of fume hoods.

Ideally, each fume hood should be individually exhausted from the building. This allows for the greatest flexibility within the hood as to selection of chemicals that can be used. It also provides the most safety in case of an accidental spill, fire, or explosion. In an individually exhausted system an accident can be contained within one fume hood, whereas, in a manifolded system the situation can spread. However, manifolded systems are more the rule than the exception. From a health and safety perspective, these systems require careful planning to avoid the use of incompatibles within the system. This requires the researchers involved with the project to develop a list of chemicals which may be used in their research. After this list is reviewed for incompatibilities, individual fume hoods need to be assigned for use with specific chemical classes. A hidden aspect to this situation is the administrative controls which the project leader must enforce in order to keep incompatibles separate.

Balancing of manifolded systems is often very difficult. The use of damper within the system was the generally accepted method until fairly recently. The use of dampers has not proven to be effective because they tend to fail for a variety of reasons and are difficult to keep adjusted. More recently, balancing of manifolded systems has been accomplished by use of static pressure differentials. This method has proved to be very effective, but has limitations.

The placement of fume hood exhaust motors is an important fire protection consideration. Fume hood exhaust motors should be placed on the roof of the building or in a fire secured penthouse. Placement of the exhaust motor directly on top of the fume hood is a fire and explosion hazard as, except for specially order motors, these motors are not sealed and are thus exposed to the chemicals they are exhausting.

Fume hood exhaust stack heights are another area of concern to health and safety specialists. Stack heights should be determined by the height of the building (building envelope), proximity to other buildings, prevailing winds, weather conditions, and location of the building's air intake. Ignoring these parameters can cause entrainment of exhaust air into the supply system, thus creating an indoor air pollution problem. As a general rule of thumb, 10 foot stack heights for single story buildings and 15 foot stacks for multi-story buildings are reasonable, provided the exhaust velocity is at least 2,500 fpm. It is important to remember that the exhaust velocity is a crucial element in the overall exhaust design. The stacks are an aesthetic problem, but the use of decorative facades can easily hide the stacks.

ANIMAL FACILITIES

Animal facilities have traditionally been under the purview of the scientist. However, there are special safety and health considerations which should be involved in the design of individual animal rooms. These facilities may, also, include housing for insects, parasites, etc.

Most often consideration is given only to keeping odors from reaching other parts of the building. From a health and safety perspective, this is the last of many reasons for the use of a separate ventilation system. In some research applications the animals in use or the diseases under study are zoonotic (animal diseases transferable to humans). Under these conditions special precautions must be taken to prevent exposure to humans. For example, sheep carry a zoonotic disease called Q Fever which is usually manifested in humans as a flu-like disease. Sheep rooms should be kept negative to the surrounding area, and exhaust air either incinerated or HEPA filtered. This disease is also transmissible to other animals such as cattle. Contamination from room to room is usually accomplished by designing each room with independent supply and exhaust ducts.

The converse situation is the housing of primates. They are extremely susceptible to human diseases such as measles and tuberculosis. It is sometimes advisable to design their holding facilities under positive pressure with limited access. A typical design is similar to containment laboratories with an ante room for clothes change and showering.

Rodents present a very different problem. They are perhaps the greatest escape artists in the world. Many experiments have been ruined because the controls and treated animals have "visited" each other or totally "disappeared." Rodent facilities need to be "escape proofed." In practical terms, this is more of a vigilant art than orderly technology. There should be as few penetrations as possible in the walls, ceiling, and floors. All penetrations should be carefully sealed, in a fashion similar to a biocontainment facility. Air vents and drains should be screened. Care must be taken not to use too small mesh as it will interfere with airflow. The wall material should be smooth.

Another problem encountered with rodents, primarily rats, is their susceptibility to respiratory diseases. Controlling temperature, humidity, and the day/night cycle are necessary to maintain the health of these animals. The answer to this problem is to incorporate of individual controls in each rodent holding area.

Insectories present another problem. Many species of insects pose allergy problems for humans. The exact nature of the allergen has not yet been characterized. However, it has been shown that continuous exposure to insect scales and fras (insect debris) can create allergic responses in sensitive individuals. Also associated with the raising of insects, are exposures to various molds, bacteria, and formaldehyde. There is no single solution to this problem, but there are good engineering controls available an specific design considerations.

Ventilation systems for insectories should be designed with directional air flow. The supply air can be directed from the front (entrance) of each room down and toward the back. Return air ducts are than placed near the floor. The supply discharge velocity should approach laminar to approximate an air curtain around workers when in the room. All surfaces should be washable, as good housekeeping is a key to allergy prevention. Often insect screening is placed over all openings. This practice is usually detrimental to the effectiveness of the ventilation system. Replaceable filters can be used which will prevent the escape of flying insects. As in the design of rodent facilities, care should be taken to seal all penetrations.

Self-contained incubation chambers are commercially available which can be used as either negative or positive pressure units. These chambers employ directional air flow so that insect scales and fras are collected on the bottom where cleaning is easier and worker exposure is minimized.

The key to design of animal facilities is simplicity. All surfaces should be washable and a water source available in each room. Each room should have individual temperature, humidity, and lighting controls. The ventilation system should be given primary consideration prior to room layouts so that the most flexibility can be designed into the facility. This becomes especially important if future research will require the separation of "clean" and "dirty" areas. Actual room layout should consider the compatibility of animal species especially with respect to cross-contamination.

In a multiple occupancy building a separate means of egress is advisable for transportation of animals to and from the building. Aside from the obvious odor containment, this egress provides protection to the animals.

FIRE SAFETY

Each individual laboratory room should have a second means of exit. Adjacent laboratory rooms may share this remote exit, via a common separation wall. The usual argument against a second exit is the scientist's need for as much wall space as possible. Although this argument is understandable it is not as important as the safe escape of workers in the event of a fire or toxic release.

The storage of flammable/combustible materials should be considered during initial laboratory design. The use of the cabinets under fume hoods, although a common practice, is not acceptable under NFPA Codes, unless the cabinets have been designed for this purpose. It is important to note that unless this type of cabinet is specifically required in the technical specifications, a typical nonflammable storage cabinet will be provided. Therefore, each laboratory should be designed to store flammable/combustible materials in a segregated, vented storage cabinet in accordance with NFPA 30 - Flammable and Combustible Liquids Code and NFPA 45.

The amount of chemicals stored in each laboratory should be limited to a short term supply (e.g., enough for one week or month). This supply by it's nature will be facility dependent. In order to allow for the storage of larger amounts of chemicals, a specifically designed area should be used. The size and building materials are specified in the OSHA Standards, NFPA 30, and NFPA 45.

Compressed gas cylinders are commonly used in laboratories. Where compressed gases are to be used which are common to several laboratories it is advisable to manifold these gases in a central location.

Sprinklers and fire protection systems are required by NFPA Codes, but are often dependent on the overall size of the facility and quantity of stored flammable/combustible material. The wisest course of action is to provide heat and smoke detectors in each laboratory and provide a sprinkler system at least in the hallways. Each laboratory should have at least one ABC portable fire extinguisher. Computers have become more important to laboratories than ever. Halon fire extinguishing systems are available which are nondestructive to both electronic equipment and human life. These should be employed for fire protection.

MISCELLANEOUS

Each laboratory should have an emergency eye/face wash and shower station. The minimum criteria for these systems are:

1. independently plumbed potable water supply
2. control valve designed to remain open without operator assistance
3. control valve to remain open until manually shut off
4. activation foot or hand treadles
5. water flow rate to meet ANSI Z358.1-81

There are numerous portable units and hand held single head eyewash devices commercially available. Some of these are good additional support, but none of them are acceptable in lieu of stationary dual head eye/face washes.

Laboratory furniture is prefabricated or custom designed for every purpose. Wood furniture is often used because of it's availability and attractiveness. The are several drawbacks to the use of wood furniture: it adds to the fire load of the building; and it is easily contaminated. In general, laboratory furniture should be constructed such that:

1. It is corrosion resistant.
2. Contamination is easily removed.
3. It can be arranged not to impede egress in an emergency.
4. The working surface is free from cracks and joints.

BIOCONTAINMENT LABORATORIES

Biocontainment laboratories are special work environments which often require special design and equipment to protect the workers and the experiments.

Until a few years ago the biocontainment level or level of protection was designated with a "P" symbol followed by a number. The "P" has been replaced with "BSL" or Biosafety Level. There are four biosafety levels which are defined according to a combination of facility design, laboratory practices and techniques, equipment and health and safety controls. It is not practical to try to completely describe all of the features and definitions pertaining to biocontainment laboratories in a chapter dedicated to an overview of design. Therefore, we will concentrate on the elements of building design for "maximum containment" or BSL-4 facilities.

A maximum containment laboratory is usually a separate building, although it can be part of another building. To maintain the required security and necessary engineering features, including ventilation and building materials, it is usually more practical to build a separate facility.

In the simplest of terms, the primary design difference between a BSL-4 laboratory and any other laboratory is the use of "secondary barriers." Secondary barriers include building materials, ventilation systems, equipment (e.g., biosafety cabinets, space suits), airlocks, change rooms, sealed openings, and decontamination systems. A BSL-4 laboratory has four "layers" between the hazardous agent and the outside environment. These layers or barriers can be achieved by using a variety of secondary barriers. There are a number of BSL-4 applications in the United States, but only one actual laboratory building. The primary considerations in deciding to use an application or build a laboratory are the hazard level of the research and the cost of the building. Although from the outside a BSL-4 building can look like any other laboratory the barriers required a quite different.

The structure must be air tight. All air from within the facility must be filtered through HEPA filters before release to the outside. Therefore, at the outset of the design process the ventilation system and the structural materials become the primary concerns. From the outer structure it is not evident that between the walls are foam materials not for insulation, but for sealing porous building materials. The use of tiling is kept to a minimum as grout is porous and allows penetration of bacteria/viruses. High quality epoxy paints are used instead as they afford the same washability and often help seal the walls. Ventilation systems are usually designed to maintain pressure differentials between different ares of the building and to provide directional airflow from the "cleanest" to the "dirtiest" areas of the building. Although more sophisticated in design and operation, these ventilation systems follow the same general principals as described previously.

Virtually everything that goes into a BSL-4 laboratory does not come out again without being sterilized, with the exception of workers. Workers are required to change clothing before entering the containment area and completely shower prior to leaving. There are some applications that require workers to shower prior to entering and again before leaving. The change area is usually located directly off of the main entrance. It should consist of a disrobe area with lockers and toilet facilities, showers, and a rerobe area. All clothing used within the containment area is sterilized between uses. Decontamination is required for all liquid effluents from within the containment area. This includes the waste from laboratory sinks, biosafety cabinets, autoclaves, toilet facilities, etc. High pressure heating vessels are usually used for treatment of liquid wastes. Even after sterilization, the processing must be tested to ensure safety prior to discharge outside of the facility. All solid waste must be incinerated or sterilized and buried.

Sometimes in the design of a BSL-4 facility, the full letter of health and safety codes/requirements for the protection of workers can not be met. This is where health and safety specialists must compromise and use their ingenuity to meet the intent of the requirements. For example, it is not always possible to provide a secondary means of egress from each area. Two change facilities are not cost effective or practical. A viable alternative is the use of airlocks with built-in liquid disinfection systems which are not hazardous to humans, but destroy the biohazard. These airlocks must be clearly identified as others are often used for transportation of equipment and other materials and contain hazardous disinfection systems.

The above elements of BSL-4 design are only the basics. Participation in the design of such a facility is extremely fascinating and difficult. Upon anticipation of such a design it is advisable to contact at least two biosafety experts who have had extensive experience in the development of maximum containment applications. The field of biosafety is rapidly growing with new applications and design criteria developing continually.

SUMMARY

The success of a laboratory design depends on many factors, not the least of which are health and safety considerations. When the team approach is implemented, each member brings to the design specific expertise essential to the element of proper design.

LITERATURE CITED

1. Occupational Safety and Health Administration Standards, Occupational Safety and Health Administration, U.S. Government Printing Office: Washington, DC, 1986.
2. National Fire Protection Association Codes, National Fire Protection Association, 1986.
3. Biosafety in Microbiological and Biomedical Laboratories, 1st Edition, U.S. Department of Health and Human Services, Public Health Service Centers for Disease Control and National Institutes of Health, U.S. Government Printing Office: Washington, DC, March, 1984.
4. American National Standards Institute Standards, American National Standards Institute, 1986.
5. Construction Policy Design Manual, U.S. Department Of Agriculture, Agricultural Research Service, 1986.
6. Laboratory Chemical Fume Hoods- Manual, U.S. Department Of Agriculture Agricultural Research Service, 1981.
7. Industrial Ventilation, 17th edition, American Conference of Governmental Industrial Hygienists: Cincinnati, OH, 1986.
8. Guide for the Care and Use of Laboratory Animals, DHEW, National Institutes of Health (out of print).
9. Health Research Laboratory Design, National Institutes of Health (out of print).
10. Midwest Plan Service, Livestock Facilities, Iowa State University: Ames, IA, 1986.
11. Comfortable Quarters for Laboratory Animals, Animal Welfare Institute (undated).
12. Chlad, F. L. Professional Safety; March, 1982.

RECEIVED March 6, 1987

Chapter 15

Design Considerations for Toxic Laboratories

William J. Maurits

Department of Army, Aberdeen Proving Ground, MD 21010-5423

Those elements of conventional laboratory design that must be refined for facilities in which toxic chemicals will be handled are presented. Alarms, communications, construction materials, containment cabinets, filter systems, floor plans, security, compressed gases, and waste disposal are discussed. Emphasis is given to design considerations dictated by the use of large numbers of fume hoods.

A successful designer of a toxic laboratory will find it necessary to refine most of the elements of the traditional chemical laboratory. Many details which aren't directly associated with the toxic operations will impact on the safety of these operations. Because common laboratory mishaps will be far more serious where toxics are used, it makes sense to invest every effort to preclude such accidents through careful design.

Floor Plan

The flow of personnel in and out of toxic areas can spread contamination, so the layout of a laboratory should facilitate routine movement of workers as well as emergency evacuations. Staff should not have to walk through one laboratory to get to another nor should an office be located where the only exit is through a laboratory. The provision of separate administrative areas will avoid locating scientist's desks in rooms where toxics are used. Visitors are safer and more easily suffered if they can view the laboratory rooms through windows.

Laboratory aisles must be no less than 5 feet wide and benches should have sufficient unobstructed width to accommodate modern analytical instrumentation. An overhead (filtered) exhaust system would permit small canopy hoods to be connected as necessary to scavenge fumes from areas near injection and exhaust ports of analyzers not located in hoods. Each room should have its own supply

This chapter not subject to U.S. copyright
Published 1987 American Chemical Society

of air for ventilation. Self closing doors will help maintain required static pressure differentials.

Emergency stations should be located near exits and should include emergency shower (with drain), and storage for blankets, towels, soap and light clothing. Eyewashes must be available in each laboratory, and should feature positive temperature control, since it is impossible to wash ones eyes for 15 minutes in icy water. Alarm pull boxes should be near each door for convenient use on the way out. Each laboratory or storeroom should have two exits (with doors that swing out) placed so that no credible event can block emergency egress. Workers must be assured an unimpeded path out of the building in the event of emergencies, so it is inappropriate to secure building doors with locks that cannot be opened from the inside. Any perimeter fencing should include gates with locks that can be opened from the inside.

Laboratory rooms intended for toxic work should be provided with adjacent shower and change facilities. The layout must not require freshly showered personnel to track back through the area that they might have just contaminated. All drains, including those in laboratory floors, should have deep traps and be directed to a toxic sump. Airlocks will help prevent toxic fumes from spreading to non-toxic areas in the event of a failure of a primary containment cabinet. Check valves in the incoming water lines will prevent contamination of potable water supplies when pressure is lost.

Secure (lockable) storage for small quantities of toxic chemicals should be available in each room. A central storage point facilitates inventorying, but must accommodate compatibility requirements for the stored items.

Primary Containment Cabinets

The nature of the work to be done, statutory requirements, and the preferences of the staff will dictate the selection of laboratory containment cabinets, but the following considerations should be taken into account by the decision makers.

Glove boxes (including Class III cabinets) may be necessary for most toxic operations or where aerosols are involved. Glove boxes permit the use of inert or otherwise controlled atmospheres. They shield the operator during use, require less ventilation than fume hoods, and don't cease to protect when house power is lost, though they may lose their negative pressure.

However, closed glove boxes are inconvenient. Materials must be passed in or out through an airlock or dunk tank and the operator is afforded only limited movement by virtue of the arm length gloves being in a fixed location. Seals and gloves will be exposed to higher concentrations of chemicals than would be generated in a hood, so organics may permeate over a period of time. Glove boxes offer less protection while seals or gloves are being changed. All work in a closed glove box is viewed through glass which seems to attract dirt on both surfaces.

Fume hoods are often selected for their convenience of use though they greatly complicate the design of a laboratory. Operators can work comfortably anywhere in the hood and materials can be brought in or out easily. Gloves can be changed conveniently without risking operator exposure to the hood's contents.

Unfortunately, fumes can drift out of a hood for a variety of reasons and aerosols will drift out. Hoods strain heating and air conditioning systems by consuming vast quantities of room air, they are incompatible with controlled atmospheres, they provide no shielding with the sash up, and their protection is degraded by turbulent flows if they are located near doors or in areas that have heavy pedestrian traffic. Flow at the hood face is obstructed by workers standing in front of the hood and all protection is lost when power failures are experienced.

The large fans associated with hoods may cause severe vibration problems unless they are appropriately mounted at some considerable distance. The mounting of blower motors behind the building can reduce unwanted vibrations in the laboratories, but care must then be taken to avoid irritating low-frequency noise from the lengthy duct work. Room air should be delivered through a perforated dropped ceiling, as it is thus more evenly delivered at lowered velocities to reduce turbulent flows.

Each hood intended for toxic work must have a face velocity of 100 linear feet per minute. When many hoods are employed, the volume of tempered air that must be supplied (summer and winter) is quite large. The required airhandling equipment is so massive that minor misadjustments may make it difficult to get out of a room because of air pressure on a door. One way to deal with this is to vent the doors and keep the hallways at a slightly higher pressure than the labs. When an airhandling (supply) unit falls short, the hallway provides needed makeup air. Computers can operate air handling systems more precisely than can traditional systems and an alarm system that pinpoints defective elements for early repair can help avoid gross imbalances.

Hoods for toxic work should be easily decontaminatable with a catch basin leading to a toxic sump. The hoods should be made of stainless steel and be conveniently locked. Provision should be made for limiting travel of the hood door to that opening which can be supported by the hood fans and the air handling system. These stops should be sturdy but adjustable.

Hoods may be required to contain considerable amounts of equipment while maintaining a specified range of air flow at the face. Therefore, the hoods must feature several internal airflow adjustments to accommodate the localized effects of equipment placed in the airpath. The hoods should be large enough to set all work back 20 centimeters or more from the face of the hood. Access through the rear panel makes the repair of contaminated equipment much safer.

Laboratories designed for the handling of toxic materials normally maintain reduced pressures in the rooms and hallways, relative to the pressure outside the buildings. Hoods should therefore be fitted with antibackflow valves to avoid sucking the contents of the ductwork into the laboratory in the event of a power failure. Backup power provided in 15 seconds does not prevent this phenomenon, even if the hoods and airhandlers are designed to restart automatically.

The floors of hoods should have lips for containing spills.

Drains should be fitted with drain plugs when not in use to ensure that toxics will not be allowed to go down the drain in an accident. It is advantageous to decontaminate toxic material before it is mixed with many gallons of diluent in a toxic sump so the toxic drain should be relied upon as a fallback sampling and treatment point.

Operations that involve transfers of toxics between containment cabinets can be conducted most safely if the cabinets are located adjacent to one another and feature interconnecting passageways. Cabinet floors can be equipped with steam baths or storage compartments to minimize the frequency with which toxic materials must be packaged up for transfer to another safe area.

The class I biological safety cabinet is intermediate between a fume hood and a closed glove box. This cabinet can be used with the front open or be fitted with gloves. Since the front access opening is normally only 8 inches high, the cabinet requires less ventilation than a fume hood. Class I cabinets are used with an airflow at the face of 100 linear feet per minute.

Filter Systems

Filter systems for toxic chemical operations usually employ a rough prefilter followed by a high efficiency particulate air (HEPA) filter, in turn followed by charcoal bed filters to remove the chemicals. Pairs of charcoal filters should be connected in series with a sampling port between filters so that breakthrough from the first filter can be detected while the excess is still being captured by the second. Influent filtering of all laboratory air is necessary to reduce the frequency with which replacement of the contaminated filters is required. Hood filter systems should be designed to reduce the hazards of change out procedures. One such system has been described.[1]

The filtered effluent from hoods must never be directed back into the laboratory. It should be released above the building at a high enough velocity to ensure that it will not be pulled into the intake vents.

Waste Disposal

All drains in a toxic laboratory with exception of those from the toilets should lead to a toxic sump. The toxic sump should be fitted with the wherewithal to permit addition of reagents, agitation, and sampling, as well as adequate indicators and alarms to highlight malfunctions. Valving should be convenient to operate and the system should feature parallel tanks so one batch can be treated while the lab continues to discharge to the other tank. Provision should be provided to pump out contents when untreatable.

The storage of solid or liquid toxic waste residues must be considered in the design of the laboratory complex. Whatever temporary storage is selected, such as berms, sheds, etc., it is imperative that a leaking drum not result in chemicals being discharged toward the aquifer. Wastes must not be stored on site for more than 90 days after collection, so the laboratory storage space may not need to be large as long as room exists for segregation of chemicals as necessary.

Compressed Gases

When such equipment as chromatographs or atomic absorption spectrophotometers are used, compressed gas tanks proliferate to the point where the quantities of energetics are too large to be safely located in a laboratory in which toxics are used. Two alternatives exist. Hydrogen can be generated electrolytically on site as needed, or it can be piped from compressed gas cylinders through manifolds. Manifolds permit the cylinders to be kept in a place more convenient to the bulk storage point and reduce the amount of such material in a toxic laboratory. The manifolds should be located where they can be checked with a soap solution regularly to find any leaks that might have developed. Manifolds should be color coded.

Construction Materials

Construction materials must be nonabsorbent and easily cleaned or decontaminated. Seamless flooring avoids cracks from which spilled chemicals can contribute a significant pollution burden to the laboratory air. Epoxy paint should be used for interior walls. Dropped ceilings should be made of nonabsorbent material such as enameled metal. Hoods and sinks should be fabricated of stainless steel. Wood or other porous surfaces must be avoided. Construction and landscaping should provide appropriate earthquake and storm resistance as well as good physical security.

Communications

Toxic operations must be supported by a good communications system. In laboratories where communications are inadequate, workers will naturally use "runners" for communication needs. This practice results in avoidable traffic in and out of toxic areas which increases the opportunities for contamination to spread. In emergencies, a phone or intercom can help ensure that assistance is tailored to the actual need. An "all purpose" response to an alarm will normally be less rapid at a time when speed may be of the essence. Video cameras trained on critical operations add a measure of safety, but annoy the workers who may feel that the purpose of the system is to "spy" on them. As a minimum, the laboratory doors should have windows so that entering personnel don't blunder into a rapidly developing scenario.

Alarm Systems

A general alarm system for a toxic laboratory should feature coded pull boxes to aid emergency response personnel in locating the specific area where the emergency exists. Sufficient audible and visible alarms should be provided to ensure that all personnel are alerted. The fact that maintenance personnel may be caught working at noisy locations above ceilings, on the roof, in service tunnels or outside the building should be considered. When activated, the alarms should continue to sound until they are turned off by human intervention. Alarm systems should be provided with reliable back up power, and should feature test circuits. Klaxons and other components should be continuous duty rated and all wiring should be encased in dedicated metal conduits.

Power surges or failures may leave the best designed airhandling equipment in various states of disarray. It is valuable in such circumstances to be able to assess the status of the hoods from outside the building. An easy to read status board can be placed so as to be visible from the outside through a window and/or remote outputs can be made available at other locations. Mechanisms for resetting the hoods should also be conveniently located.

The exhaust duct of each ventilated containment cabinet must be fitted with an adjustable low flow sensor. Audible and visible alarms must be located near the cabinet, and the silence switch should energize an indicator at the status board. These are local alarms which should not automatically trigger a call for emergency response personnel.

An alarm system should be provided to warn workers of power interruptions that have occurred during non-duty hours. Such evidence that engineering controls have been compromised alerts incoming personnel to the necessity for first entry monitoring of laboratory rooms.

Power

Ground fault interrupters should be included in all circuits used to power laboratory instrumentation. Circuit breakers should be near the areas they serve. Emergency lighting must be provided in each room, hallway and staircase. It is common practice to utilize battery powered lights for this purpose. House power is used to keep the batteries charged.

Security

The use of toxics carries with it a responsibility to maintain an effective system to ensure that dangerous chemicals are not released to unauthorized persons. The entire building should be within a secure perimeter and/or individual laboratories or suites of laboratories should be securable. Within laboratories and stockrooms there should be secure storage for any toxics and other controlled substances that are used.

Security systems are available featuring magnetic badges, personnel identification numbers, passphrases, or even digital or retinal scanners that unlock those specific areas to which the individual employee has been granted access. Since these systems are computer controlled, the access authorization for any individual can be conveniently and quickly adjusted as circumstances warrant. Logging of traffic in the various areas can be accomplished automatically. It should be understood that computerized systems are susceptible to intrusion and may therefore lack the positive control of a well organized and monitored system of secure keys or combinations.

Literature Cited

1. Barberto, M. S. In Toxic Chemical and Explosive Facilities; Scott, R. A., Ed.; ACS Symposium Series No. 96; American Chemical Society: Washington, DC, 1979; pp 192-214.

RECEIVED May 13, 1987

Chapter 16

Design of Blast-Containment Rooms for Toxic Chemical Ammunition Disposal

Paul M. LaHoud

Huntsville Division, U.S. Army Corps of Engineers, P.O. Box 1600, Huntsville, AL 35807

Environmentally safe destruction of obsolete chemical weapons must be performed in facilities which assure total containment of blast effects and toxic gas in the event of an accidental detonation. Functional process requirements and recommended structural design procedures for containment rooms to accomplish this purpose are presented. The requirements presented are consistent with Department of the Army and Department of Defense Explosive Safety Board requirements.

A variety of chemical warfare (CW) munitions have been manufactured by the United States ending in the late 1960's. Large quantities of these CW munitions remain stored at several U.S. Army installations. The CW agents contained in these munitions are extremely toxic compounds that produce lethal or incapacitating effects on man. The two general categories of concern are nerve agent and mustard-blister agents. The nerve agents are organophosphate chemicals. The mustard-blister agents, also called vesicants, are systemic poisons.

A wide variety of weapon configurations were designed to dispense these agents. These included bombs, rockets, mines, spray-tanks, cartridges, mortars and projectiles. The U.S. stockpile of these munitions ranges from 18 to 32 years old. The agent contained in the munitions is even older and has begun to deteriorate in storage. In many cases, weapon systems to deliver these munitions are no longer in service. Many of these munitions pose an additional hazard resulting from the presence of explosive bursters, fuses and propellant. None of these munitions were designed to facilitate disassembly at the end of their useful life. Figure 1 illustrates a typical explosively configured weapon.

Rising concern over the deterioration of these munitions in storage and the related safety and environmental risks, led to Public Law 99-145, which directs the Secretary of Defense to carry

This chapter not subject to U.S. copyright
Published 1987 American Chemical Society

out the destruction of the U.S. stockpile of CW munitions by September 30, 1994. Responsibility for implementation of the requirements of this law rests with the Office of the Program Manager for Chemical Munitions (OPMCM), Aberdeen Proving Ground, Maryland. The Huntsville Division of the U.S. Army Corps of Engineers, Huntsville, Alabama is providing engineering and contracting in support of the execution of this program.

Functional Process Requirements

The Army terminology for destruction of obsolete weapons is "demilitarization". This term encompasses all the steps required to disassemble and safely destroy or decontaminate the component materials of which the munition was constructed. National Academy of Sciences and Department of the Army Guidance for demilitarization of obsolete chemical weapons (1) requires absolute safety and security, assurance of total containment of agent during processing, maximum protection of operating personnel and incontrovertible evidence verifying the destruction of the toxic wastes.

The functional steps in the destruction of explosive chemical munitions include:

1. Safe disassembly of the munition and removal of the explosive components and propellant.
2. Disposal of the explosive components and propellant.
3. Accessing the agent cavity of the munition.
4. Disposal of the CW agent.
5. Disposal of the munition bodies.
6. Disposal of the process generated waste streams.

The approved method for disposal of chemical agent and decontamination of other munition components is incineration (2). Figure 2 presents the functional disposal process selected for this program.

The dominant process criteria is agent containment. Overall containment within the process facility is accomplished by maintenance of negative pressures within the building. The negative pressures increase progressively as ventilation air passes from low risk areas into higher risk areas. All ventilation air is "once through" and then treated using high efficiency charcoal filters before release to the environment. Assurance of agent containment in areas where explosives are removed from munitions requires total blast and fragment containment and the capability to confine the residual toxic gas products in the event of an accidental detonation during processing.

Explosive Containment Requirements

The design requirements for the explosive containment rooms in the facility are defined using the detailed process operating requirements and safety and environmental factors:

1. Total containment of blast and fragmentation effects in the event of a detonation.

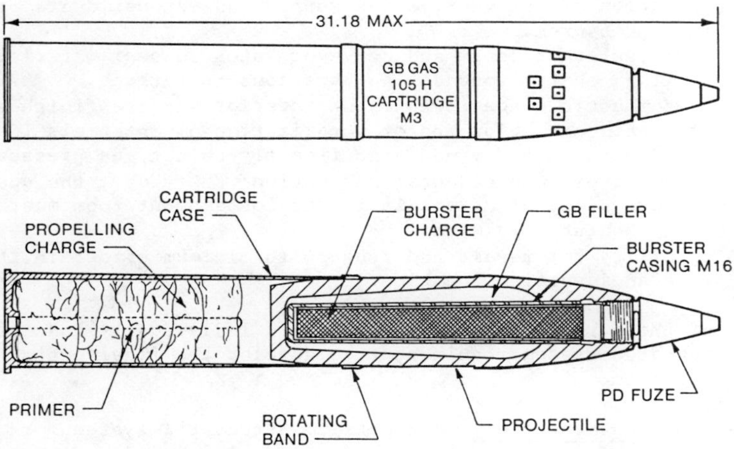

FIGURE 1: CARTRIDGE, 105 MILLIMETER: AGENT GB, M360

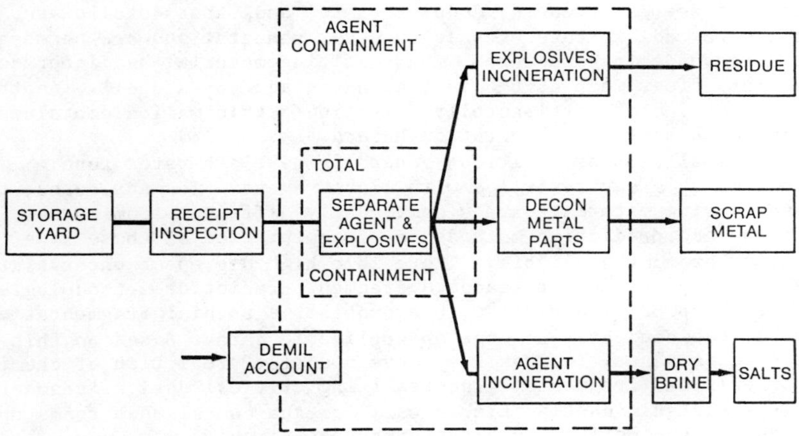

FIGURE 2: CHEMICAL MUNITION DEMILITARIZATION SCHEMATIC

2. Total containment of post-detonation toxic hot gas products until safe for processing.
3. Protection of the ventilation supply and exhaust ducts from blast pressures.
4. Blast resistant doors and conveyor gates to seal material handling penetrations during hazardous operations.
5. Non-combustible agent-resistant interior surface finishes. Note that combustion and/or vaporization of materials in the containment room may add significantly to hot gas pressures in the event of an accidental detonation; therefore, the quantity of these kinds of materials in the containment room must be kept to an absolute minimium.
6. Capability for repair and reuse with minimum effort in the event of an accidental detonation.

Each of these requirements is considered individually and then as an integrated system requirement to develop the final containment configuration.

Blast and Fragmentation. The optimum structural system for confinement of explosive shock and residual gas pressures would intuitively appear to be some form of a shell of revolution such as a sphere, or cylinder with hemispherical heads. A structural material such as steel with good tensile strength can be used with great efficiency in this fashion. However, as the total system requirement is considered, this initial economy is rapidly eroded by other factors. Stiffeners, doubler plates and other details are required to redistribute stesses whenever penetrations are necessary in a stressed skin structure. The resulting material and labor cost penalties offset much of the initial advantage for a shell. Another significant factor detrimentally affecting a thin walled containment was found to be the fragmentation hazard.

Chemical weapons munitions generally have a burster tube surrounded by a cavity filled with liquid agent. In many cases, the burster casing materials are significantly different from normal munitions and prediction methods for fragmentation of these type munitions are not available. There is a high degree of uncertainty regarding application of standard fragment prediction methodologies to these weapons. To resolve this problem, a special fragmentation test (3) was conducted to develop applicable data. Based on this test data, a manual (4) was then developed for prediction of chemical weapon critical fragments. The resulting critical design fragment requires a significantly thicker wall for the containment rooms than is required to confine the blast pressures alone.

The final element which influenced the room shape selection was volumetric efficiency. To provide a given room floor area and overhead clearance requires a much larger volume for a shell of revolution than is required by a more typical rectangular-shaped room. The unusable extra floor space and volume to be ventilated in a spherical or cylindrical shell are significant penalties. The results of this evaluation lead to the conclusion that a rectangular-based cubicle is the preferred room configuration. Additional parameter studies concluded that in the rectangular cubicle configuration, reinforced concrete is the preferred construction material over structural steel. Design of reinforced concrete structures to resist blast forces is based on well proven procedures (5). Recent experimental data from a model structure similar in configuration was also available to validate the design methods. A detailed discussion and design example of this model is presented elsewhere in this Handbook under the title of "Structural Design for Blast Containment."

Containment of Gas Pressure. In the event of an accidental explosion during munition disassembly, the highly toxic agent in the munition would be released. The total containment criteria dictates that any such release be confined in the process facility containment room. The energy released by the explosion would vaporize the agent and heat the air in the room to a high temperature. Because the air cannot be vented, a substantial gas pressure will develop and exist after the blast shock waves have dissipated. The containment room must safely confine this pressure until it decays through heat transfer to the surrounding concrete. As the gas cools the internal pressure will decrease until it reaches a level suitable for processing through the ventilation system.

In practice, total containment is difficult to achieve since there will be some leakage around door seals, conveyor gate seals and through the concrete itself. Consideration was given to providing a vapor tight liner plate to minimize risk of leakage through the concrete. Such a liner plate would have to be sufficiently thick to assure that no fragment penetration occurred. In addition the liner plate would have to be erected in segments, seal welded and then have concrete cast against it. The practical difficulties in accomplishing these actions reliably are significant. In addition, there was concern that voids could exist between the liner and the concrete. Leaks in welds could allow agent migration into these voids, and these dangerous pockets of contamination would be undetectable. It was preferred that the concrete be exposed to allow verifiable decontamination if required.

To assure confinement in the facility of the total leakage from all possible sources, the explosive containment rooms are surrounded by a plenum area which is maintained at negative pressure. The ventilation rate of this plenum area is designed to easily accommodate the projected leakage from the containment room after an incident. Live explosive model tests (6) were used to predict vapor leakage through the concrete. The rate of leakage is a direct function of the internal pressure after an incident. Testing confirmed that the confined gas cools rapidly, with proportional decrease in internal pressure. Thus, the leakage rate also decreases at the same rate. Figure 3 presents graphically this mechanism. Information shown in the figure is closely representative of the expected performance of the actual design. Pneumatic pressure testing will be performed after construction to verify design leak rates are not exceeded.

Ventilation System Blast Protection. The explosive containment rooms have the highest potential contamination level in the process facility. The punching and shearing that are part of the remote controlled disassembly operation result in the release of significant agent vapor in the rooms. For this reason the containment rooms are maintained at the highest negative pressure in the facility and a high rate of air change is maintained continuously. All ventilation air passing through the containment rooms leaves the facility and goes directly to the filters. It is critical that the containment rooms have the capability to quickly isolate the ventilation supply and exhaust ducts in the event of an explosion. This isolation is achieved by providing a quick response blast-actuated valve in series with a controllable gas tight valve for both the supply and exhaust ducts. The blast-actuated valve provides protection from the explosive shock pressures and the gas valve provides positive gas leakage control thereafter.

Figure 4 shows the final ventilation system protection scheme. It should be noted that even with a blast valve that closes in a few milliseconds there will be some reduced shock pulse that "leaks through" during closure of the valve. The peak value of this shock is a function of losses occurring as the shock passes through the valve and the duration is the valve closure time. The leakage shock was predicted using the blast valve manufacturers' test data. Figures 5a and 5b, respectively, show representative values for the incident shock and the leakage shock passing the blast valve. This loading was then used to analyze the ventilation ducting to assure no damage would occur.

Blast Resistant Penetrations. All doors, conveyor penetrations, feed chutes and utility penetrations must be designed to assure the total containment requirement is not compromised. They must be operationally reliable and well sealed to minimize leakage to the plenum area surrounding the containment rooms. Design of these elements revealed that the fragmentation threat was the governing factor and required 2.5-inch steel plate. Obviously doors and conveyor gates made of plate this size required powered operators. Compression seals were also used for leak tightness. The door, conveyor gates and feed chute doors are remotely controlled by the process control system. These assemblies are factory tested to assure that they operate and meet the minimum leak rate requirements. Frames for these closures are cast into the concrete at the time of construction.

Surface Finish Materials. The explosive containment rooms will be exposed to a harsh environment during the lifetime of the facility. The toxic agent exposure level is high. The surface coating system for walls, roof and floor must be non-reactive and impermeable to these exposures. Decontamination during maintenance or equipment changeout will require room washdown with highly caustic decontamination solutions. The surface coating system must also survive in this environment. An epoxy coating system has been tested and approved which does not absorb or react with the chemical agents and is functionally resistant to the washdown solutions. A secondary benefit of the surface coating system is its sealing of the concrete which improves its vapor tightness.

The presence of the coating system as well as other materials which were potentially combustible raised the risk of causing additional increases in the post-detonation gas pressure. Recent experimental work (7) has confirmed the significance of this

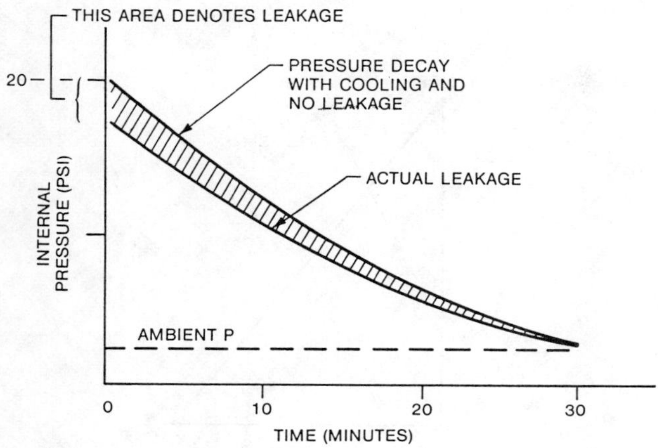

FIGURE 3: INTERNAL PRESSURE DECAY WITH TIME

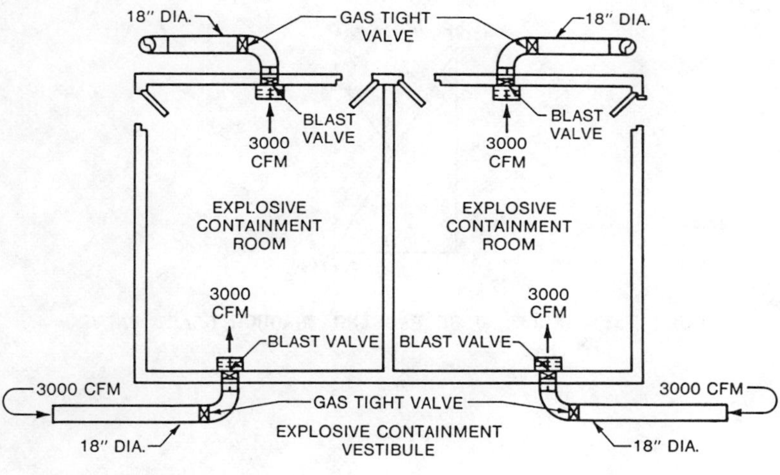

FIGURE 4: EXPLOSIVE CONTAINMENT ROOM VENTILATION SYSTEM BLAST PROTECTION

a

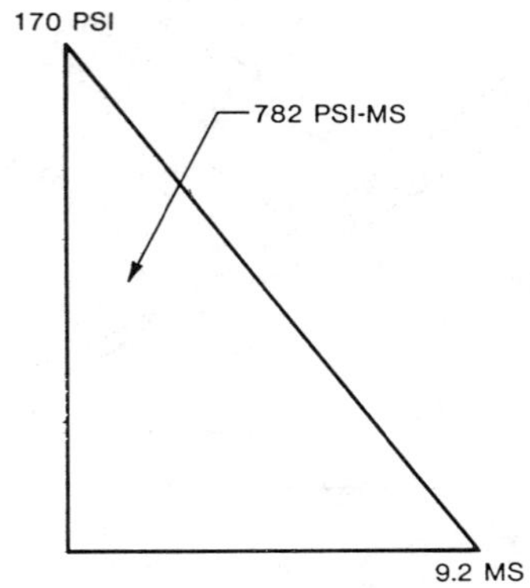

FIGURE 5a: SHOCK PULSE AT INLET SIDE OF BLAST VALVE

b

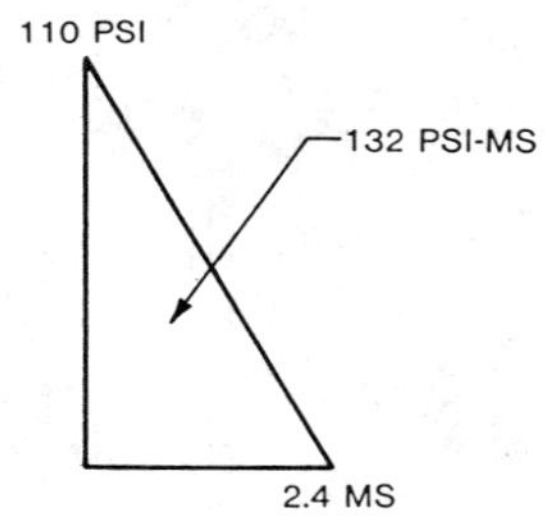

FIGURE 5b: SHOCK PULSE PASSING THROUGH BLAST VALVE

a

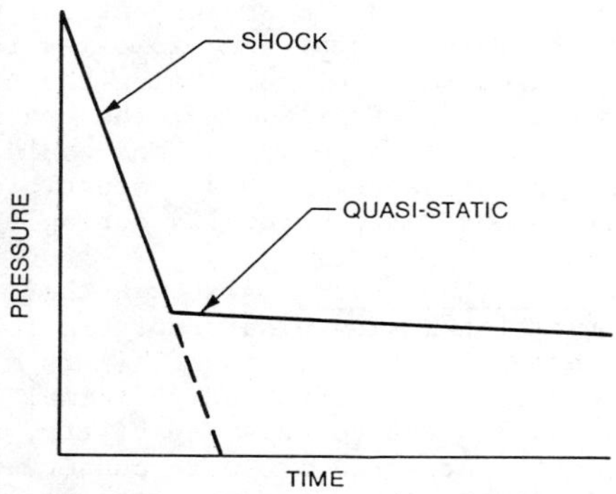

FIGURE 6a: EXPECTED PRESSURE-TIME HISTORY FOR DETONATION IN CONTAINMENT ROOM WITH NO COMBUSTIBLES PRESENT

b

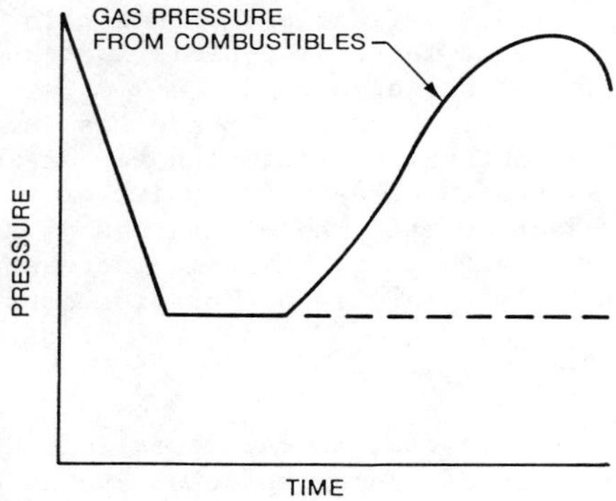

FIGURE 6b: EXPECTED PRESSURE-TIME HISTORY FOR DETONATION IN CONTAINMENT ROOM WITH THE BURNING OF COMBUSTIBLES

phenomena. Figure 6a shows the expected pressure time history for a detonation in a containment room. Figure 6b shows a similar event except the burning of combustible materials present in the test caused a dramatic increase in the subsequent gas pressure. To assure no such risks were present, an explosive test program (8) was conducted on a model containment room using the proposed surface coating system. This test verified that the coating was not combustible for the conditions expected and would not, therefore, contribute to the gas pressure. Other combustibles expected to be in the rooms will be monitored carefully during operations.

Repair and Reuse After Explosion. Although the risk of a high order detonation of a munition during disassembly is low, this hazard does exist. In the event of such an incident, it is a design requirement for the containment rooms to suffer only minimal damage and allow rapid refurbishment. To assure this capability, the containment room structural design criteria are more conservative than Department of Defense Explosive Safety Criteria would normally require. This is considered appropriate since vapor containment is so critical in this facility.

During the transient load phase of an accidental explosion, when the shock duration is less than the time of maximum response of the structural elements, member end rotations are limited to one degree. Maximum inelastic deformation is limited to three times the member elastic limit deflection. Since this loading phase is suddenly applied, use of material dynamic increase factors based on strain rate of loading are also used.

After the transient shock load phase has damped out, the subsequent confined hot gas pressure can be considered as a steady state load from a structural dynamics point of view. Therefore the design criteria requires that these loadings do not exceed the elastic limit of the structure. Dynamic increase factors are not applicable since loading rate is no longer a consideration.

Summary

Integration of explosive containment rooms into a process facility requires consideration of overall process system performance not simply the structural design elements. Use of reinforced concrete for containment design is a viable and economical choice of material for the facility requirements of this process. Design procedures for reinforced concrete subjected to blast loads are well documented and tested and are suitable for containment design. Additional considerations are present in containment structure design which are neglected during design of vented structures. These include long term gas pressure, additional pressures from combustion products and validity of material allowables and deformation limits. Safety dictates that these elements be considered carefully.

Literature Cited

1. Report of the Disposal Hazards of Certain Chemical Warfare Agents and Munitions, prepared by an Ad Hoc Advisory Committee of the National Academy of Sciences, Washington, DC, June 24, 1969.

2. Technical Paper Defining the Operating Conditions for the Incineration of the Chemical Agents GB, H, and VX, U.S. Army Toxic and Hazardous Materials Agency, 7 May 1984.
3. Powell, J.G. Fragmentation Characterization Profile for Chemical Filled Munitions-M23 Land Mine, 115MM Rocket Warheads and 8-inch Projectiles, Naval Surface Weapons Center, April 1983.
4. Whitney M.G.; Friesenhahn, G.J.; Baker, W.E.; Vargas, L.M. A Manual To Predict Blast and Fragment Loadings from Accidental Explosions of Chemical Munitions Inside An Explosive Containment Structure, U.S. Army Toxic and Hazardous Materials Agency, April 1983.
5. Structures to Resist the Effects of Accidental Explosions, U.S. Army Technical Manual, TM 5-1300, Washington, DC, June 1969.
6. Kirtland Underground Munitions Storage Complex Model Designs, Construction, and Test Data, Technical Report SL-84-14, U.S. Army Waterways Experiment Station, prepared for the Defense Nuclear Agency, Washington, DC, April 1984.
7. Hokanson, J.C.; Esparza, E.D.; Baker, W.E.; Sandoval, N.R.; Anderson, C.E. Determination of Blast Loads in the Damaged Weapons Facility, Volume 1, Final Report for Phase II, Southwest Research Institute Project 06-6578, Pantex Plant, July 1982.
8. JACADS Explosive Containment Room Model Test, Southwest Research Institute Project 06-8069, prepared for the U.S. Army Corps of Engineers, Huntsville Division, Huntsville, Alabama, July 1984.

RECEIVED March 6, 1987

OTHER DESIGN CONSIDERATIONS

Chapter 17

Intrinsically Safe Electrical Circuits in Explosives Facilities

Kenneth W. Proper

U.S. Army Defense Ammunition Center & School, Equipment Division, Savanna, IL 61074-9639

During design of explosive facilities, one of the major concerns is limiting the electrical energy which can ignite the often present, hazardous environment due to sparks or thermal effects. Intrinsically Safe Circuits provide a means of accomplishing this. However, the successful utilization of intrinsically safe electrical circuits depends upon a complete understanding of not only its construction requirements, but also its concept. Therefore, in order to provide this understanding, a presentation of its history, definition, application, and general construction requirements are presented. More importantly, its virtues and disadvantages are discussed.

In the design of explosive facilities, two major considerations are of paramount importance: controlling the conditions which can lead to a premature initiation of energetic materials, and providing the maximum degree of personnel and property protection.

Controlling the conditions which can lead to a premature initiation of energetic materials can be accomplished through the elimination of energy sources within the hazardous environment. However, in doing so, the capability to accomplish the mission is also eliminated. Therefore, the goal is to provide the amount of energy which will accomplish the mission; yet, do so in such a way as not to provide energy which can cause initiation of the energetic material.

One method of limiting the amount of energy capable of causing initiation has been through the use of pneumatic and

This chapter not subject to U.S. copyright
Published 1987 American Chemical Society

hydraulic equipment. However, the major disadvantages of this approach are the complicated logic systems required and the slow response times, especially for sensing and metering equipment.

Another approach has been through the use of explosion-proof electrical installations to provide the energy required to accomplish the mission. This method does not limit the amount of energy, rather its philosophy is if an explosion occurs, to contain that explosion within its heavy wall construction and prevent its propagation to the outside environment.

An additional approach, which permits the use of electrical energy in the hazardous environment, is the use of purged and pressurized enclosures for electrical equipment. Once again, this approach rather than limiting the energy depends on not allowing the hazardous environment to come in contact with the electrical energy, thereby eliminating the probability of an explosion. This is accomplished through purging of the electrical installations and maintenance of a positive pressure in them so that the environment within the electrical system is non-hazardous.

A final approach, which permits the use of electrical energy within a hazardous environment, is through the use of intrinsically safe electrical circuits. Rather than restricting the propagation of an explosion or maintaining a non-hazardous environment, it reduces the amount of electrical energy within the hazardous environment to levels which are incapable of igniting that environment. This concept is not new; due to new advances in technology, its application has greatly increased in scope.

History of Intrinsically Safe Electrical Circuits

At the turn of the century in Germany, research was begun on the effect of an electrical spark on methane-air mixtures. This work would play an important role several years later in Britain.

In Britain in 1912 and 1913, a rash of mine explosions lead to a formal court inquiry. It was found that at this time the practice of signaling was accomplished by the rubbing together of two bare wires connected to a battery to form a circuit. As a result of the court findings, testing became required for signaling equipment in British mines.

This task was assigned to what is now called the Safety in Mines Research Establishment. It was at this organization where the concept of intrinsically safe electrical circuits was first defined after continued research into the ignition of methane-air mixtures.

In 1936, the first certificate was issued in Great Britain for an intrinsically safe electrical device which was not designed for application in mining operations.

In 1938, the United States Bureau of Mines began development of rules relating to the use of electricity for telephone and signaling equipment, which included application of intrinsically safe electrical circuits.

Until the 1950's, the use of intrinsically safe electrical circuits had little application in other than just battery operated signaling devices. At this time due to advances in technology and an increase in the use of electrical equipment in hazardous locations, a new world-wide interest developed in the application of intrinsically safe electrical circuits beyond what had been its traditional role.

In the United States, this new interest was recognized, and in 1956, the National Electrical Code (NEC) introduced the use of intrinsically safe electrical circuits.

> "Equipment and associated wiring approved as intrinsically safe may be installed in any hazardous location for which it is approved, and the provisions of Article 500 and 510 will not apply to such installations."(1)

However, no guide was given for the construction nor testing of the circuits.

In 1967, the National Fire Protection Association (NFPA) issued NFPA 493-1967 which defined specific tests and construction techniques to be employed. Today, the current standard is NFPA 493-1978.

World-Wide Acceptance

Intrinsically safe electrical circuits are now recognized around the world as an additional technique for providing electrical energy in hazardous locations.

However, the standard used in the United States and the standards used in Europe do not coincide. The dissimilarities are due to a difference in the manner in which hazardous environments are classified and to a divergence in philosophy over the safety factor employed. NFPA 493 uses a safety of 1.5 pertaining to the total energy, while the International Electrotechnical Commission (IEC) and European Committee for Electrotechnical Standardization (CENELEC) require a safety factor of 1.5 for the voltage or current, which relates to a 2.25 factor of safety for the energy.

What Are Intrinsically Safe Electrical Circuits?

Definition: Webster's defines intrinsic as "naturally, essentially, or inherently" and further defines safe as "free from damage, danger, or injury; unable to cause trouble or damage" (2). From these definitions, a definition of intrinsically safe can be derived to mean: inherently and naturally unable to cause trouble, damage, or injury.

Due to this derived definition, circuits are mistakenly considered as intrinsically safe due to the circuit utilizing low energy. However, in reality, the circuit may not qualify as intrinsically safe because the definition as stated in NFPA 493-1978 qualifies the above definition.

> "Intrinsically Safe Circuits: A circuit which any spark or thermal effect, produced either normally or in specified fault conditions, is incapable, under the test conditions prescribed in this standard, of causing ignition of a mixture of flammable or combustible material in air in its most easily ignited concentration."(3)

The qualification being that it must fail safe not only in its normal mode of operation, but, also, under specific modes of failure. Therefore, it is not enough to state that the circuit is of low voltage, and because of this is intrinsically safe. This is only half of the requirement. To qualify as intrinsically safe, the circuit must also fail in such a way as to be incapable of causing ignition, and further, it must be either tested or analyzed according to prescribe methods.

Evaluation of Intrinsically Safe Circuits

NFPA 493-1978 is very explicit in Chapter 2 as to basic requirements which must be met in order for a circuit to be considered intrinsically safe. They are:

1. The normal operation shall not be capable of igniting the hazardous environment when adjusted for its worst operating conditions and an additional energy factor of 1.5 is introduced;
2. The circuit must be incapable of igniting the hazardous environment when operated at 1.5 its energy rating and the inducement of one fault and its related failures. Further, the circuit must be incapable of igniting the hazardous environment at its normal energy rating when two faults and their associated failures are introduced;
3. Intrinsically safe circuits shall conform to the construction requirements contained in Chapter 3 and 4 of the standard.

Defining the Hazardous Environment

The first task, which should be completed before considering the design of any facility or equipment involving energetic material, is to define exactly what type of hazardous environment will be involved in each room, section or area. This is a prerequisite, whether selecting intrinsically safe or any other technique to provide electrical protection.

The classification of hazardous locations involves the determination of four factors:

1. What are the hazardous elements in the process?
2. Are the hazardous elements vapors or dusts?
3. What are the explosive and/or electrical characteristics of the hazardous elements?
4. Are the hazardous elements constantly present or only present under special circumstances?

The Hazardous Element. To often it is automatically assumed that in an explosive facility the explosive item is the most hazardous item and, therefore, the electrical protection is designed based on its requirements. However, this assumption can lead to installing the wrong type of electrical protection.

All the processes being performed, in each room, section, or area, must be carefully reviewed to determine if other elements being used pose a greater hazard. The following questions can serve as a guide in reviewing the processes to determine all the hazardous elements involved:

1. Are elements given off which are more hazardous, i.e. gases from chemical reactions?
2. Are elements introduced into the process which are more hazardous, i.e. large volumes of flammable solvents during rework processes?

Vapor or Dust. Once the hazardous elements for each process within the facility have been identified, it is necessary to determine whether they constitute a hazard due to being a vapor or a dust. Vapor and dust represent two different types of explosion hazards.

Explosions from vapors occur due to the rapid transfer of heat from one molecule to the next molecule. Additionally, vapors can only ignite when present in certain concentration ranges - known as their lower and upper explosive limits. Also, vapors disperse due to diffusion and convection; therefore, if a vapor cloud is released and is not ignited, the hazard is soon gone.

Dust presents a different type of hazard, because while it has a lower explosive limit, it does not have an upper explosive limit. This can result in a primary explosion, followed by secondary explosions as new air is provided. Secondly, dust does not diffuse away from its point of release, but settles out of the air and accumulates into layers. Unlike vapor, the dust explosion is caused by the radiant heat from one particle igniting the next. Because of this, the lower explosive limits for dusts are greatly higher than for vapors. Also, the size and shape of the dust particles are important factors in effecting its lower explosive limit.

Due to the differences in behavioral characteristics, different approaches are used to prevent their accidental ignition due to the presence of electrical energy. The National Electrical Code (NEC) recognizes three classes of hazardous environment. Class I being for hazardous environments consisting of flammable vapors or gases; Class II for hazardous environments resulting from the presence of combustible dusts; and Class III for fibers and flyings, usually associated with the textile industry.

It is important to note that each class employs a different type of philosophy to prevent ignition. Therefore, Class I rated protection may not provide protection when used in Class II or Class III environments or vice versa.

Type of Vapor or Dust: The NEC further subdivides Class I and II into groups. Groups A through D are used to denote groups of equivalent types of gases or vapors present. While Groups E and G are used to denote groups of equivalent types of dust hazards based on their conductivity. Group F is used to denote carbonaceous dusts.

Likelihood of Hazard. The NEC recognizes two distinct levels of hazard probability. Division 1 denotes an environment in which the probability exists that sufficient levels of the hazardous element may always exist, under normal operating condition, as to warrant extreme protections. Whereas, Division 2 denotes an environment where the probability for sufficient levels of the hazardous element to exist, under normal operating conditions, is less likely, and therefore, the extreme protection is not justifiable. Further areas adjacent to Division 1 areas can often constitute classification as Division 2 environments.

The Explosives' Environment. The Army Materiel Command (AMC), which has the primary responsiblity for manufacture and storage of explosives for the Department of Defense, clarified its definition of the type of hazardous location involved with explosives, propellants, and pyrotechnics in its most recently revised safety manual (4). When the only consideration for hazardous environment is the presence of explosive material, it recommends that the environment be classified as Class II, Group G, with the appropriate division based on the probability of the hazardous element being present in the environment. It further states that consideration must be given to vapors which might be present or to the presence of metallic dust.

NOTE

For complete definitions and classification of hazardous electrical environments, consult Article 500 of the NEC.

Completing the Evaluation. Once the hazardous environment has been classified, the design of the electrical protection can be completed. It may require only fulfilling the requirements for one class and group, or several groups within one class, or even two classes and several groups. Whatever the result, the cost of the installation can be greatly reduced by this action while ensuring the maximum degree of protection is being provided. This is possible since equipment can be selected which was designed to fulfill the requirements explicitly for that environment, rather than a wide spectrum of requirements for all possible hazardous environments.

Intrinsically Safe and the Explosive Environment. If the evaluation concludes that the environment is in fact a Class II, Group G, Division 1 location, then the intrinsically safe electrical circuits must be designed as dust-tight and meet the requirements for Class I, Group D as defined in NFPA 493. For other types of hazardous environments, the intrinsically safe electrical circuits must be designed to meet the requirements of NFPA 493 for that type of environment.

How it works

Intrinsically safe electrical circuits in a sense are usually composed of two different circuits. One of which is located in the hazardous area, while the second is located in the non-hazardous area. The former being a low energy circuit connected to a metering, sensing, or an enabling device, while the latter being connected to a controlling, indicating, or instrumentation device.

Electrical Isolation. These two circuits are integrated to create one circuit through a safety barrier. The purpose of this safety barrier or protective interface is to ensure electrical isolation so that the higher levels of energy available in the non-hazardous circuits cannot be transmitted to and through the circuits in the hazardous area.

Circuits Not Device. During design, when considering the use of intrinsically safe electrical circuits, the whole electrical circuit must be considered. It is not enough just to consider the electrical apparatus employed in the hazardous environment. Consideration must be given to its associated apparatus located in the non-hazardous area. Therefore, it is not just the apparatuses which must be considered, but the whole circuit, both in the hazardous area and the non-hazardous area.

Safety Barriers. Figure 1 illustrates an application employing intrinsically safe electrical circuits for the demilitarization of ammunition. Three separate areas are required for this application - one area, classified as non-hazardous, to serve as the control and loading area; a second area, classified as hazardous, where the actual demilitarization is accomplished; and a third area, classified as non-hazardous, is required for the hydraulic pump due to the level of noise produced.

The hazardous area was classified as Class II, Group G, Division 1 due to the projectiles being separated from the cartridges and the propellant being dumped into a vacuum collection system. The operation of the machine's pneumatic and hydraulic systems are controlled and verified by the use of intrinsically safe electrical circuits. The control cabinet located in the non-hazardous area consists of a programmable controllor, other electrical equipment, and safety barriers. All signals passed to or received from the hazardous area by the controllor are conducted through safety barriers. This ensures that any faults occurring in the non-intrinsically safe circuits could not result in dangerous energy levels being passed to the hazardous location.

For this application, Zener safety barriers were selected as the protective interface. Further, every circuit going into the hazardous area is connected to a separate Zener safety barrier.

Zener safety barriers are probably the most widely used and acceptable method of limiting the energy (5). A Zener safety barrier consists of Zener diodes and resistors in a network. The resistors limit the current and protect the diodes, while the diodes limit the voltage and allow grounding of the circuit. The working rating of the Zener diodes is chosen to be above the peak value of the normal working voltage of the circuit. Several companies manufacture modular forms (6) which offer flexibility of design and at the same time are tested and approved for use.

However, other type of protective devices are available which can be used. They are:

1. Transformers, three different types are discussed in NFPA 493;
2. Current-limiting resistors;
3. Blocking capacitors;
4. Shunt diodes;
5. Relays;
6. Self contained apparatus, i.e. battery operated.

The construction requirements which must be met by each of the above are contained in Chapter 3 of the standard.

Physical Separation. In addition to providing electrical isolation, it is necessary to provide physical separation to ensure the non-hazardous circuits can not degrade the intrinsically safe portion of the circuits. This can be accomplished by planning the physical layout to incorporate the use of distance, enclosures, partitions, separate raceways, and insulation. The final physical layout selected should meet or exceed the requirements of Chapter 3, Sections 1, 2, 3, and 4 of NFPA 493.

Additional Requirements. In addition to electrical and physical isolation requirements, the surface temperature of all equipment and wiring located in the hazardous environment must not exceed the values indicated in the standard.

Further, the apparatus must be marked according to the requirements of Section 2 of Chapter 4 of the standard.

Demonstration of Requirements. The use of electrical, physical separation is demonstrated in Figure 2. The safety barriers are contained in a separated compartment within the electrical control cabinet. Each of the safety barriers are positioned so that the intrinsically safe terminals are facing each other (Figure 2). This allows easier segregation of the non-intrinsically safe wires from the intrinsically safe wire. For added protection, the wiring is enclosed in grounded, metal raceways for support and additional isolation. Each safety barrier is grounded, and this common ground is earthed

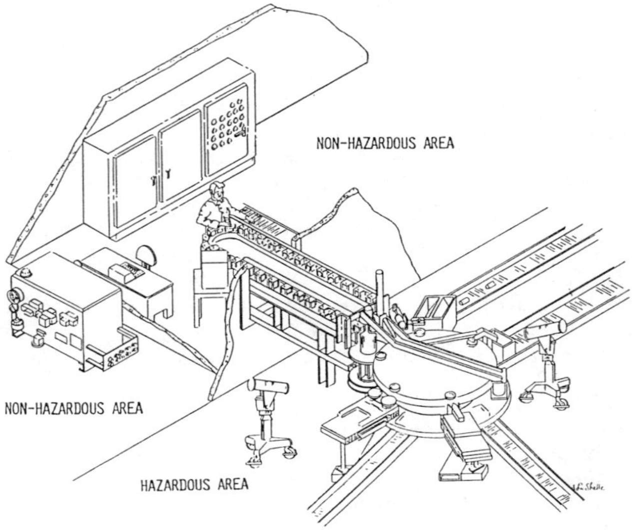

Figure 1. Application employing intrinsically safe electrical circuits.

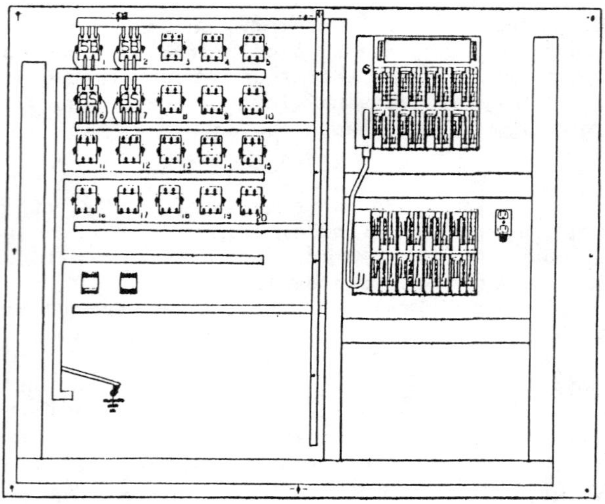

Figure 2. Electrical and physical separation.

separately from non-intrinsically safe circuits (Figure 2). To provide additional protection, the safety barriers are bright blue in color and marked as required. Following a local requirement, blue tape is wrapped around intrinsically safe wiring every few inches for easy recognition.

Demanding Requirements

The design requirements for intrinsically safe would seem to be demanding, and a review of NFPA 493 enforces this fact. Today's industrial environment imposes additional requirements not only on the use of intrinsically safe electrical circuits, but other hazardous electrical techniques as well. These requirements are due to the Occupational Safety and Health Act and the employer's increasing vulnerability for liability.

OSHA Requirements. OSHA requirements state that one of three options must be fulfilled regarding the selection of electrical equipment for locations classified as hazardous (7).

The first option permits the selection of instrinsically safe equipment and associated wiring. The equipment and wiring must be approved for the hazardous location in which it will be used.

The second option permits selection of approved equipment. However, not only must it be approved for the hazardous class, but, also, for the specific type of vapor, dust, or fiber involved.

The last option allows the employer to select equipment which is safe for the hazardous location. While the equipment does not have to be approved, the employer must be able to demonstrate that the design will provide the protection necessary to prevent the ignition of the vapors, liquids, gases, dusts or fibers in the hazardous location.

Employer Liability. Today more than ever before, employers are being challenged by their employees to prove that all possible effort was employed to reduce hazards in their work place. Many employers had not been able to prove they had done this, and, therefore, they have suffered costly settlements and increased liability insurance expenses.

Certification. It is a benefit to the employer to ensure that the intrinsically safe electrical circuit is certified. Certification can be achieved through the use of a third party, such as Underwriters Laboratories or Factory Mutual Research. Both of these organizations have their own standards for approval which are based on NFPA 493. The certification is accomplished in three steps:

1. The circuit is analyzed to determine faults and operating characteristics;
2. The circuit is reviewed to ensure construction and temperature requirements are met;

3. The performance characteristics are verified by either the actual testing of the circuit in its intended environment or comparing calculated or acutal measured values against the graphs in Chapter 5 of the standard.

Benefits of Intrinsically Safe Electrical Circuits

In spite of the rigorous design requirements and the need for certification, instrinsically safe electrical circuits offer many advantages which the other hazardous location electrical techniques do not.

First, once designed, evaluated, and installed, the safety of the system cannot easily be degraded because the safety is in the design, not protection added afterward. In fact, the intrinsically safe electrical circuit will cease to fulfill the function for which it was designed long before it can become a hazard. This is due to the consideration which must be given to fault conditions. The only possible way for the circuit to become hazardous is if an unapproved or unauthorized component is substituted into the circuit.

Secondly, the circuits do not require the additional expenditure of money for added protection to ensure the safety of the designed system as do other techniques used for hazardous wiring.

Thirdly, the cost and time for installation is less, again due to safety being in the design and not added protection, which must carefully be installed to ensure it provides the degree of safety required.

Fourthly, intrinsically safe electrical circuits are the easiest to maintain. Since intrinsically safe circuits by their nature are incapable of causing ignition, they can be maintained without regard to shutting down operations, nor are hot permits required, or is lengthly disassembly, assembly and recertification of added protection required.

Finally, due to the requirements for intrinsically safe circuits being the most conservative of hazardous location circuits requirements, intrinsically safe electrical circuits offer the maximum in safety. Not only do they control the conditions which can lead to initiation of energetic materials, by their very nature - they eliminate it.

Intrinsically Safe Circuits The Easy Way

The simplest method of using intrinsically safe electrical circuits is not to design and certify them yourself, but rather to take advantage of a clause contained in NFPA 493 which states:

> "One of the serious problems which has faced both manufacturers and users in applying the intrinsic safety concept has been the inability to interconnect apparatus of different manufacturers and

> be assured that the combination is still intrinsically safe. The marking scheme below ....(explains the marking system and requirements)... The above (marking system) information and cable characteristics are all that are necessary to determine that independently certified intrinsically safe and associated apparatus may be interconnected, without loss of intrinsic safety. It should be recognized that this procedure results in systems which are evaluated with as many as four independent faults."(8)

Through the use of this clause, the design time can be reduced and the problem of certification can be eliminated. It now becomes a matter of defining the problem to be solved, defining the environment, selecting commercially available equipment which is rated as compatible for the task, following the manufacturer's instruction for installation and verifying the cable characteristics.

Availability. Both Underwriters Labatories and Factory Mutual Research publish yearly guides to electrical equipment which they have certified and continue to certify as being rated for use in hazardous environments. Many of the items contained in these guides are rated as intrinsically safe or as associated equipment for use with intrinsically safe equipment. Further, the amount of equipment available should increase each year as the demand increases for intrinsically safe electrical circuits.

Real World Application

Due to the concept of low energy, intrinsically safe electrical circuits do not provide the energy necessary to drive motors or high powered electrical equipment. Nevertheless, this does not limit or restrict their application in the real world.

As mentioned earlier, pneumatic and hydraulic systems have been extensively used in hazardous environments to provide the power necessary to move and drive machinery to complete needed tasks. Their use has demanded development of complex logic systems which involve the addition of valves and piping. These logic control systems are often hard to design, debug, construct, and maintain.

The advent of the programmable controllor has allowed complex logic systems to be easily developed and permits greater control over processes than ever before. They can interpret both digital and analog signals. They are capable of multi-tasking, permitting one unit to control several different processes at the same time. They can be connected to main frame computers, enabling process data to be centrally collected for both coordination of processes and report generation.

Another new important tool for use in hazardous locations is robotics. This tool allows the operator to be removed from the hazardous environment to a location away from the danger, affording the operator maximum safety.

Intrinsically safe electrical circuits provide the capability to combine the strengths of pneumatic and hydraulic systems with the sophistication of the programmable controllor and robotics, and to do so with the maximum safety and flexibility.

The U.S. Army Defense Ammunition Center and School is employing the use of intrinsically safe electrical circuits in equipment designed to demilitarize and renovate munitions - from small arms to large projectiles. This is accomplished by using pneumatics and hydraulics to provide the power, while using position switches and solenoid valves linked to programmable controllors to direct the total machine process.

In one application, the same programmable controllor coordinating the actions of the machine is also providing control over an intrinsically safe robotic arm which loads and unloads the heavy projectiles being processed. In this way, the maximum protection is afforded not only to the operator, but also to the facility.

Conclusion

The utilization of intrinsically safe electrical circuits when possible during the design of explosive facilities, can accomplish one of the paramount objectives - controlling the conditions which can lead to a premature initiation of energetic materials in the environment. This is possible because intrinsically safe electrical circuits are designed to be incapable of igniting the hazardous environment, not only when operating correctly, but even when malfunctioning.

The ideal way to accomplish this utilization is through purchasing certified apparatuses and combining them to arrive at the circuit desired, rather than designing the apparatus and circuit. This simplifies the design process and, further, provides documentation for OSHA requirements.

Finally, intrinsically safe electrical circuits are an old idea, whose time has just begun. Tomorrow's world will see ever greater uses of programmable controllors, robotics, solid state circuits, and other low energy devices. This is the world in which intrinsically safe circuits belong.

Acknowledgments

The author wishes to acknowledge the assistance of Ms. Flanagan, Ms. Miatke, Mrs. LaShelle and the USADACS Equipment Division.

Literature Cited

1. Article 500, National Electrical Code, National Fire Protection Association, Quincy, MA, 1956

2. Webster's Twentieth Century Dictionary, 2nd Ed., Unabridged, William Collins + World Publishing Co., Inc.
3. Section 1-4.1. Reprinted with permission from NFPA 493-1978, Standard for Intrinsically Safe Apparatus and Associated Apparatus for Use in Class I, II, and III, Division 1 Hazardous Locations, Copyright 1979, National Fire Protection Association, Quincy, MA 02269. This reprinted material is not the complete and offical position of the NFPA on the referenced subject, which is represented only by the standard in its entirety.
4. Chaper 6, "Electrical Equipment and Wiring", AMC Regulation 385-100, U.S. Army Materiel Command, 1984
5. Garside, Robin, Intrinsically Safe Instrumentation, A Guide, North American Edition, Instrument Society of American, Research Triangle Park, NC, 1978, has good discussion of Zener barriers.
6. Garside, Robin, "Modular Zener Barrier Design Simplifies IS Installations", page 45, Control and Instrumentation, Vol II, No. 1, Jan 1979
7. 1910.307(b)(1)-(3), 29 Code of Federal Regulations, Parts 1900 to 1910, Revised as of July 1, 1984, Office of the Federal Register, National Archives and Records Service, General Services Administration, U.S. Government Printing Office: Washington, DC, 1984
8. Appendix A-4.2. Reprinted with permission from NFPA 493-1978, Standard for Intrinsically Safe Apparatus and Associated Apparatus for Use in Class I, II and III, Division 1 Hazardous Locations, Copyright 1978, National Fire Protection Association, Quincy, MA 02269. This reprinted material is not the complete and offical position of the NFPA on the referenced subject, which is represented only by the standard in its entirety.

Other Standards

Manufacture, Transportation, Storage, and Use of Explosive Materials, NFPA 495, National Fire Protection Association, Quincy, MA, 1985
Purged and Pressurized Enclosures for Electrical Equipment, NFPA 496, National Fire Protection Association, Quincy, MA, 1982
Approval Standard Intrinsically Safe Apparatus and Associated Apparatus for Use in Class I, II and III, Division 1, Hazardous Locations, Class No. 3610, Factory Mutual Research, Norwood, MA, Oct. 1979
Intrinsically Safe Apparatus and Associated Apparatus for Use in Class I, II and III, Division 1, Hazardous Locations, UL 913, Underwriters Laboratories, Inc., Northbrook, IL, July 1979
Installation of Intrinsically Safe Instrument Systems in Class I Hazardous Locations, ISA-RP12.6, Instrument Society of America, Research Triangle Park, NC

Other References

Bartkus, Albert A., "Intrinsically Safe Equipment For Use in Class I, Division 1 Hazardous Locations", Lab Data, Winter/Spring, 1982

Capp, B. and Widginton, D.W., "The Intrinsic Safety of Resistive Circuits", 2nd International Conference on Electrical Safety in Hazardous Environments, 9-11 Dec 1975, pages 43-47 Electrical Equipment for Hazardous Locations, Underwriters Laboratories, Inc., Northbrook, IL

Magison, Ernest C., Electrical Instruments in Hazardous Locations, 3rd Ed., Instrument Society of America, Research Triangle Park, 1978

Magison, Ernest C., Intrinsic Safety, Instrument Society of America, Research Triangle Park, 1984

Redding, R.J., "The Use of Solid State Circuitry Within Hazardous Areas", 3rd International Conference on Electrical Safety in Hazardous Environments, 1-3 Dec 1982, pages 219-233

Weatherhead, D., "Intrinsic Safety", Measurement and Control, Vol 10 No. 9, Sept 1977, pages 341-349

Zborovszky, Zsuzsanna and Louise A. Cotugino, A Comprehensive Study of Intrinsic Safety Criteria, Bureau of Mines, Open File Report 23-73, March 1972

RECEIVED March 6, 1987

Chapter 18

Electrostatic Studies in Army Ammunition Plants

William O. Seals[1], James Hokenson[2], and George Petino[3]

[1]Army Research, Development and Engineering Center, Dover, NJ 07801-5001
[2]Southwest Research Institute, San Antonio, TX 78284
[3]Hazards Research Corporation, Denville, NJ 07834

One of the greatest hazards that exist in the manufacture of solid propellants, explosives, and pyrotechnic materials is dust explosions. At the different stages of manufacture, considerable quantities of dust can be produced. These unwanted quantities of dust are produced during the screening, drilling and packaging operations. In addition to posing a fire/explosive hazard, health problems for plant personnel can be serious. It is essential that the dust be removed safely from each operation. To accomplish this removal, exhaust fans are used to extract dust from the surrounding atmosphere and deposit it in transport ducts. The dust is then air carried through the ducts to a dry dust collector or passed through a water blanket for removal. The collision of dust particles with each other and the frictional forces upon each particle as it contacts the air can produce hazardous levels of electrostatic energy. Dusts which do not contain an oxidizer have an upper explosive limit. When these dust concentrations are sufficiently high enough, the fuel-air ratio of the cloud can produce an energetic reaction; therefore, dust concentration levels under dynamic flow in a dust collection system were desirable. The interrelations of duct size, dust concentration levels, and flow conditions that can produce hazardous initiating and propagating reactions within the ducts needed to be addressed.

This chapter will discuss the evaluation of dust explosion potential at various manufacturing operations in three Army Ammunition Plants. The assessment of data from each plant will be presented in detail.

Army Ammunition Plant Dust Evaluation

Three Army Ammunition Plants were selected to evaluate whether dust explosions could occur in their explosive materials manufacturing operations:

0097-6156/87/0345-0269$06.00/0
© 1987 American Chemical Society

1. Louisiana AAP, Shreveport, La.

2. Longhorn AAP, Marshall, Texas

3. Lone Star AAP, Texarkana, Texas

In each of these plants, the characterization of the dust explosion potential was carried out by sampling transport ducts for explosive dust concentrations during an actual plant operation. The critical measurements taken were the quantification of explosive dust concentrations and level of electric energy generated from the electrostatic charge accumulations found in the duct.

In order to characterize the concentration of dust flowing inside a duct, a measured amount of dust must be extracted over a known period of time. This collection velocity must be the same as the internal duct flow velocity to avoid altering the distribution of dust particle sizes. In addition, a number of sample points over the entire duct cross sectional area is necessary to define the overall dust concentration. This method of sampling, known as gravimetric sampling under isokinetic conditions, was used to determine the dust concentrations at the various manufacturing areas in the Army Ammunition Plants.

Duct Velocity and Flow Rate

To measure the internal flow velocity in the duct, dust sampling was taken at various points along the vertical diameter. A pitot static tube and magnehelic gauge, shown in Figure 1, was the equipment used for these measurements. The duct humidity, tempertaure, and static pressure were measured to calculate the gas density. In determining the humidity, the wet and dry bulb temperature of a continuous sample stream was used. To prevent dust buildup on the wet bulb thermometer, an inline metal filter was inserted into the line.

Dust Concentration

Dust samples were collected by the probe/filter configuration shown in Figure 2. The filter used to trap the explosive dust was a 37mm plastic filter cassette. To monitor the actual flow rate, a rotometer was used. The calculation for each traverse point dust concentration was obtained from

$$C_i = \frac{W_{Di}}{Q_{si}\, t_{si}} Q_{si}\, t_{si}$$

where:

C_1 = dust concentration in the duct.
W_{Di} = weight of dust collected on filter cassette.
Q_{si} = probe sample flow rate.
t_{si} = sampling time.

Note: subscript i = value at the ith traverse point.

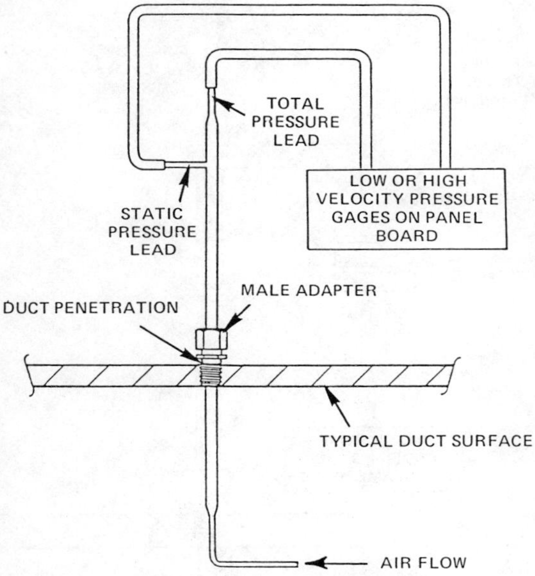

Figure 1. Pitot-static velocity probe.

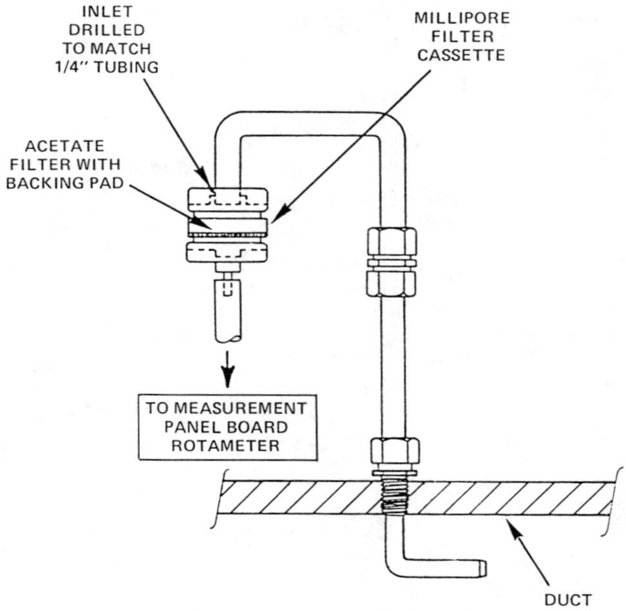

Figure 2. Dust sampling probe.

Electrostatic Instrumentation

The charge density of dust transported through ducts and the resultant electric fields at the duct inner walls was monitored by a Monroe Electronics Inc., Model 171 electric fieldmeter. All the electrostatic sampling in the field was performed in circular cross-section ducts. Thus, the electrostatic field intensity, for this geometry, can be determined from Poisson's equation using the cylindrical coordinate system.

Calibrations

The Monroe Electric Field Meter was calibrated by using a voltage standard and a large parallel plate capacitor. The electric field between the two parallel plates is calculated as a function of voltage across the plates. The calculated field is used to determine the calibration constants. To calibrate the charge density meter, simultaneous electrostatic measurements are made using the charge density meter and electric field meter. By comparing the simultaneous measurements under uniform space charge conditions, the transfer function for the charge density meter was determined from the electric field meter as the standard. The transfer function accounts for flow conditions, effects of the medium being measured, and the characteristics of the sampling hose. The transfer function determined was based upon Composition B explosive dust flowing through 305m (100 ft.) of 2.54cm (1 in.) diameter conductive hose at 9.4 l/s (20 cfm)

$$\rho = 36.9 \left[\frac{100}{G}\right] V_o \quad n\ C/m^3$$

where C = gain of charge density instrument

V = output voltage

$n\ C/m^3 = 1.0 \times 10^{-9}$ coulombs

Charge Density Measurements

A charge density meter, shown in Figure 3, designed and built by Southwest Research Institute was used to record the charge density measurements. This meter consisted of a sensor unit, control readout unit, and power supply. Basically, this instrument operates by extracting a dust sample from a duct and then passing through the sensor unit. Here, a series of steel screens trap the charge laden dust particles. To avoid hazardous charge buildups in the sensor, the charge is removed from the steel screens to ground. This creates a current flow that can be converted to voltages. It is this voltage that is recorded.

Plant Sampling and Results

Louisiana AAP

Two different process areas were selected at the Louisiana AAP for dust concentration and electrostatic charge accumulation determination. These areas were (1) the Composition B screening and bin loading in building 1611 and (2) the 155mm shell drilling operation in building 1619.

Building 1611

Bulk Composition B explosive is received in 27.4 kg (60 lb) boxes and conveyed to the second floor. The explosive is dumped on a shaker and screened to remove foreign matter. In this operation, a considerable amount of dust is generated. The dust is contained by vented hoods above the shaker and transferred into 30.5 cm (12.0 in.) ducts. The screened material then drops through a duct to a loading hopper on the first floor. The explosive dust generated by this process is removed through a 10.2 cm (4.0 in) duct. The 12 inch and 4 inch ducts are connected in a Y configuration that leads into a 12 inch duct to a wet collector. This collection system is shown in Figure 4. The cleanout openings in the ducts that facilitate the removal of dust accumulations were used as the sample collection areas. To record the dust velocity, probes were installed in the duct. One of the most essential features of this probe was its round bottom which prevented disturbances in the flow during normal operations.

Building 1619

The drilling operation, which provides a recess for the installation of a fuze in a 155mm shell, was performed in building 1619. An air driven drill is used to put a recess in the Composition B that has been encased in the nose. The dust generated from this operation is removed by suction through a 5.1 cm (2.0 in) line to a Hoffman primary dust collector. Downstream of the primary collector is a secondary collector used to take any excess not trapped in the primary collector. Two sample areas were selected for study as shown in Figure 5.

Dust Concentration Measurements

In both building locations, the velocity profile indicated duct floe turbulence. The drilling operation of building 1619 had flow velocities and negative static pressures that were significantly higher than the operations in building 1611. These differences can be attributed to the duct diameters, sizes, and number of dust cleanouts found in the two removal systems.

Sampling of the dust concentration was made at the centerline and one point above and one point below the centerline. A close inspection of the data indicated that a higher dust concentration was observed at the bottom of the duct with essentially constant levels from the top of the duct to the centerline.

Dust concentrations were three orders of magnitude higher for the drilling operation in 1619 than obtained in the hopper loading operation of 1611. This was to be anticipated when one analyzed the two types of activity. It had been found that the drilling of 48 shells would accumulate 11.34 kg (25 lbs) of explosive dust.

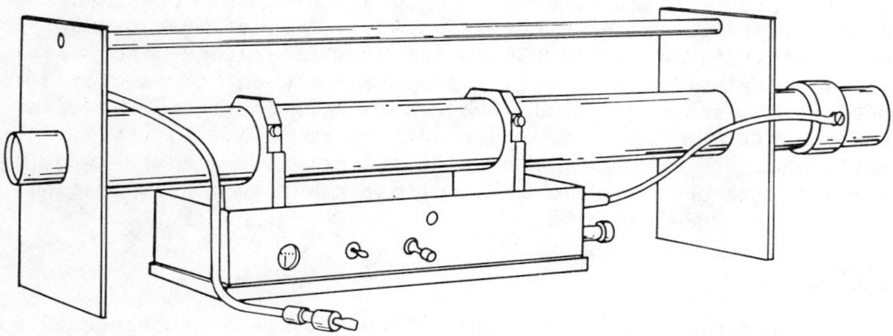

Figure 3. Charge density sensor.

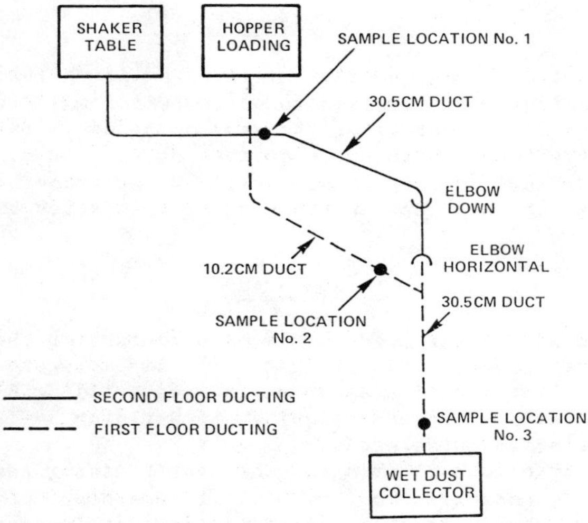

Figure 4. Dust and electrostatic sampling location in the Composition B screening and bin loading operation of Building 1611, Louisiana AAP.

Electrostatic Measurements

Building 1611

Electric field and charge density measurements were recorded at each sample location in building 1611. Typical measurements are shown in the Figure 6. In the strip chart recordings, each peak in the electric field traces, corresponds to when Composition B was dumped on the shaker. The lag corresponds to the length of time taken for the dust to be transported through 30.5m (100 ft.) of sampling hose. In spite of this delay, one can see that there is excellant agreement between the two instruments for the duration of each pulse and arrival time.

Building 1619

The duct diameters were 5.1 cm (2.0 in); thus, instrumentation was limited to the charge density meter for collecting data. Shallow drilling charge density measurements were made at locations 4 and 5 in Figure 5. The magnitude of the charge at either of these points showed no significant differences. Since the charge density signal was dependent upon the operator, no predicable characteristics could be rendered from one signal to another from the random loadings.

Charge and Energy Levels

Although the charge density levels in building 1619 are two orders of magnitude greater than found in building 1611, the energy levels are all approximately of the same magnitude. This is based upon the energy level dependent upon the duct diameter. The levels levels of energies found at these locations were many orders of magnitude smaller than the reported ignition energies for Composition B.

Longhorn AAP

Longhorn AAP is involved in the manufacuture of the 4.2 illuminating flares. Two sites, buildings B-7 and 34Y, were selected for dust and electrostatic measurements. In building B-7, 4.2 aluminum candles are processed; while, in Building 34-Y, white signal flares are manufactured.

Processing of 4.2 illuminate consists of mixing the composition, weighing, consolidation, removal of a carboard plug, adding a primer stage, and packaging. A schematic of this processing operation in Building B-7 is illustrated in Figure 7. The same manufacturing process steps followed in building B-7 are found in Building 34-Y. The sampling areas for Building 34-Y are shown in Figure 8.

Duct Velocity, Flow Rates, and Dust Concentration Measurements

The processes monitored were not continuous; therefore, the

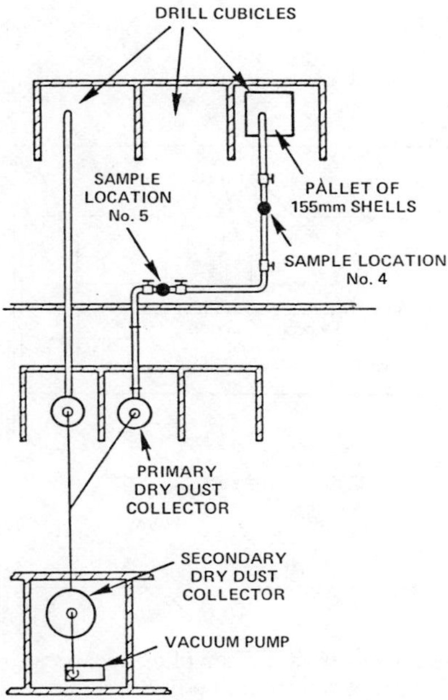

Figure 5. Dust and electrostatic sample locations in the drilling operation of Building 1619, Louisiana AAP.

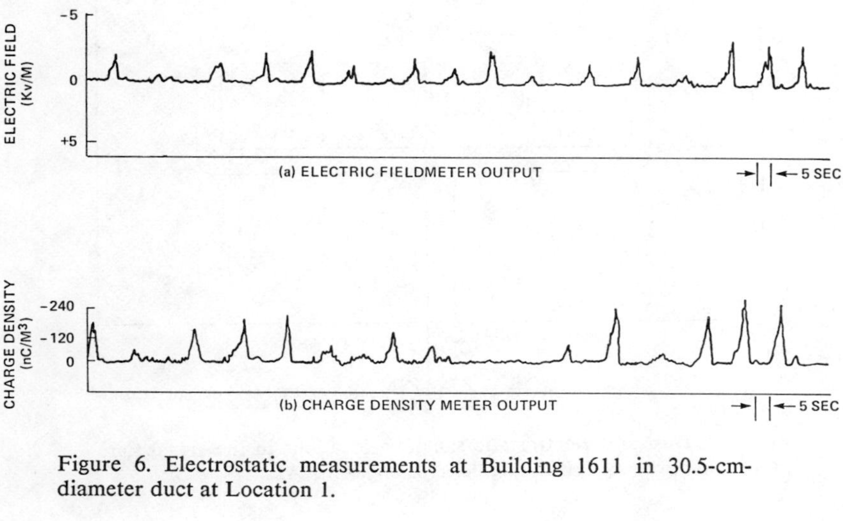

Figure 6. Electrostatic measurements at Building 1611 in 30.5-cm-diameter duct at Location 1.

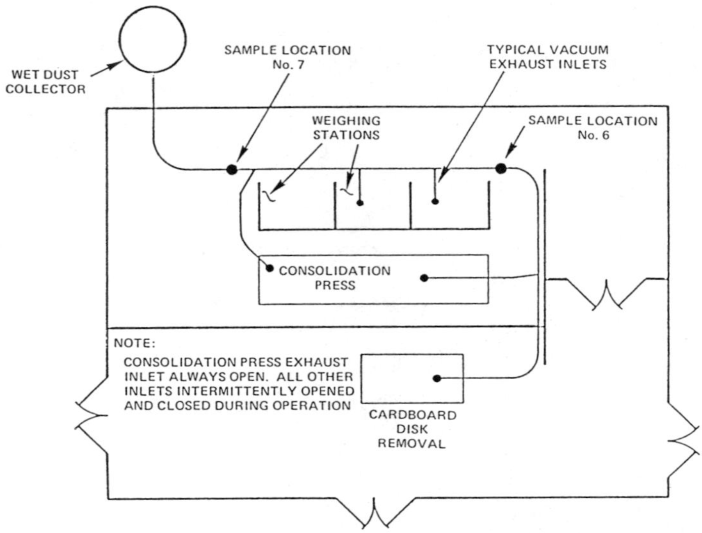

Figure 7. Dust and electrostatic sampling locations in 4.2 aluminum candle production process in Building B-7, Longhorn AAP.

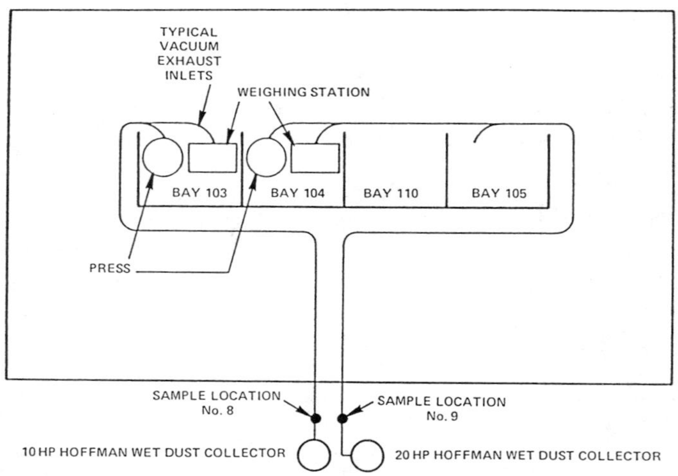

Figure 8. Dust and electrostatic sampling locations in the signal flare production process in Building 34-Y, Longhorn AAP.

consistency in the measured values was poor. This was attributed to intermittent vacuuming performed at the discretion of the operator. Only the inlets on the consolidation press had its dust vacuumed continuously.

Electrostatic Measurements

The small 2.0 in. ducts in buildings B-7 and 34-Y limited the instrumentation studies to the charge density meter. The same locations cited for dust velocity and flow rate sampling points were used for these measurements. At this location, pyrotechnic materials are processed. These materials differ from the Composition B that was used in the original calibration of the charge density meter. As a consequence of not using the pyrotechnic material with the electric fieldmeter to calibrate the charge density meter, only relative charge levels can be inferred from the data.

While these small diameter ducts produced high charge levels, the energy levels in the transport system were small. Positive and negatively charge species were found to co-exist. The positive charges occured from the intermittent vacuum at the weigh station and the negative charges from the continuous vacuuming at the consolidation presses.

Building B-7

A typical charge density waveform from the sample 6 location reflects the dust taken during the vacuum operation at the disk removal station. As seen in Figure 9, the charge can be either positive or negative. Typical polarity charge reversals can be attributed to the transfer of image charges.

Building 34-Y

Sampling points in Building 34-Y were selected near two wet collectors of two independent vacuum collection systems. It was interesting to note that the dust collected at these points were granular and larger in size than dusts collected at any of the other plants analyzed. Apparently there is sufficient moisture or volatile content to cause the fine magnesium and aluminum particles to agglomerate into large particles. The charge magnitudes were observed to be higher in the morning. As the temperature increased in the afternoon, this charge magnitude was seen to decrease. Moisture condensate formed on the duct surfaces as the temperature changed. These moisture and temperature variations may have contributed to the decreased charge levels.

The dust from the weighing and pressing stations of Bay 103 were sampled at location 8. Again, the sampling of dust was performed by the operator in a random fashion. This random operation produced unpredictable charge density waveforms. The charge density levels are quite high, but the energy levels are low. These low levels are attributed to the small duct diameters and dependency of the energy upon the duct radius to the fifth power. The energy levels at building 34-Y are approximately an order of magnitude lower than those observed at building B-7. This lower order was due to the agglomeration of the aluminum composition that occurred in building 34-Y.

Lonestar AAP

The burster facing operation (building 04-M-40) and a grenade pressing operation (building B-46) were sampled at Lonestar AAP for dust and electrostatics. These operations were similar in nature as those performed in Building 1619 at Louisiana AAP and the pressing operation at Longhorn AAP.

The vacuum exhaust and dust collection system is illustrated for buildings 04-M-40 and B-46 in Figure 10 and 11 respectively. In building B-46, three separate operations are performed: consolidation, demachining, and cone swagging. A rotary press is used to consolidate A-5 explosive into a grenade. To pick up the dust from this pressing operation, flexible rubber hoses (2.0 in.) are used. These lines are then connected to a stainless steel line that runs into a wet collector.

Dust Measurements

The flow rates and static processes are approximately the same for all vacuum lines. The velocity profile does show turbulence in both processes. Except for the drilling operation in the Louisiana AAP, the dust concentrations at location 10 and 11 were the highest recorded. In location 10, the dust concentration was more concentrated at the bottom, while the top and centerline concentrations were fairly uniform.

Of the three operations in building B-46, higher dust concentrations were generated by the demachining operation. Fairly constant concentrations were found across the duct. This can be attributed to the high duct flow velocities.

Electrostatic Measurements

The operations studied were limited to the charge density meter because of the small ducts. The dust collected from the rotary drill and facing machine at location 10 had the highest charge levels measured in the entire testing program. It soon became apparent in the initial start up of the sampling that the steadily increasing charge levels would exceed the measurement range of the charge density meter. At this point, the flow rates through the charge density meter were reduced from 9.4 /s (20 cfm) to 7.1 /s (15 cfm). The peak measurements for the two flow rates were than compared. The charge density meter transfer function at a flow rate of 7.1 /s (15 cfm) was found to be

$$\rho = 156.8 \left[\frac{100}{G}\right] V_o \ nC/m^3$$

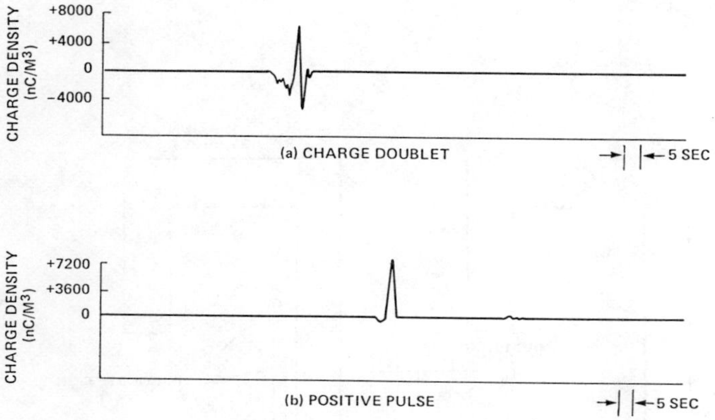

Figure 9. Charge density measurements at Building B-7 at Sample Location 6.

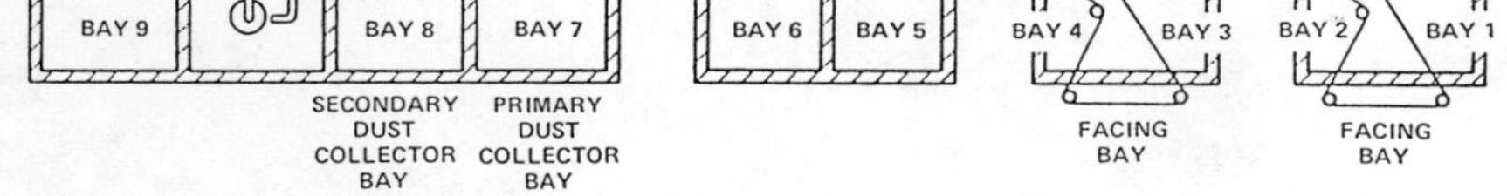

Figure 10. Vacuum exhaust ducting and dust collection system for burster facing operation in Building 04-M-40.

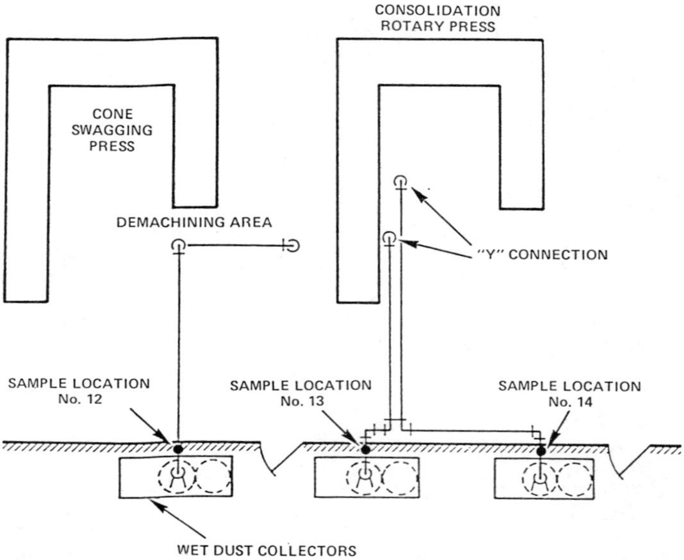

Figure 11. Vacuum exhaust and dust collection system for grenade press operation in Building B-46.

Building B-46:

Distinct and unusual waveforms were observed from A-5 explosive dust collected at location 14 when the explosive material is dumped from a bucket into a rotary press hopper. Positive and negative charge species were found with the predominance of charge being negative in polarity. Distinct charge doublets result each time a bucket is emptied. With the deposition of a negative charge in the press hopper, the opposite image charge is retained by the powder remaining in the bucket. As the bucket is completely emptied, the negative charge reaches its maximum and then begins to diminish. As a result of this action, the charge reverses its polarity . This phenomenon is completed when the image charge doublet of the opposite polarity is formed and returns to zero when the bucket is empty.

Charge and Energy Levels

Building 04-M-40 recorded the highest density levels of any of the sample locations measured. In addition, the highest readings were also always obtained when the sample was withdrawn from the bottom of the duct.

Summary of Plant Sampling

A summarization of all the data collected at the three Army Ammunition Plants is given in Table 1. The maximum values obtained at each sample location have been listed in this table. Although the results from the different processes are difficult to compare, these qualitative observations can be made.

- Sampling in small diameter vacuum ducts resulted in higher vacuum pressures, flow velocities, dust concentrations and charge densities, but lower flow rates.

- Higher charge densities, dust concentrations, and energy levels were found in processes involving drilling, and facing operations of explosive.

- Low flow velocities prevented uniform dust concentrations in the ducts. (This was reflected in the dust buildup at duct cleanouts).

- Batch operations have periods of high and low loading densities. This indicates that the gravimetric method of sampling, dependent on the total mass of dust collected over a given period, can only reflect average concentrations. Instantaneous concentrations may be significantly higher.

- Minimum explosive concentration for explosive and pyrotechnic dusts have been reported* in the range of 40 to 1000 gm/mm^3, (40 to 1000 oz/ft^3). With the exception of location 5 in Building 1619 at Louisiana AAP, all the dust concentrations determined for the various plants were below the maximum average concentrations.

Table 1. Summary of Measurements Taken During the Plant Sampling*

SAMPLING LOCATION	MATERIAL SAMPLED	DUCT-DIA (CM)	FLOW RATE (M^3/MIN)	STATIC PRESSURE (MM Hg)	TEMP (°F)	REL HUMIDITY (%)	DUST CONCENTRATION (GM/M^3)	CHANGE DENSITY (nC/M^3)	ENERGY (pJ)
LOUISIANA AAP									
BLDG 1611: 1	COMP B	30.5	63.00	−2.29	62	41	0.093	−232	2.430
2	COMP B	10.2	4.90	−2.29	63	50	1.610	+184	0.005
3	COMP B	30.5	68.00	−2.29	72	36	0.115	−287	3.000
BLDG 1619: 4	COMP B	5.1	—	—	—	—	—	+14,800	1.020
5	COMP B	5.1	3.80	−152.40	75	35	330.000	+11,100	0.570
LONGHORN AAP									
BLDG B-7: 6	ALUMINATE	5.1	0.47	−127.00	88	6	0.900	+7,750	0.280
7	ALUMINATE	5.1	4.90	−108.00	89	42	6.300	−11,100	0.570
BLDG 34-Y: 8	ALUMINATE	5.1	3.00	−88.90	89	36	12.100	+3,500	0.057
9	ALUMINATE	5.1	4.20	−101.60	89	50	1.400	+1,030	0.005
LONESTAR AAP									
BUILDING 04-M-40: 10	COMP B	5.1	3.30	−50.80	79	67	26.000	140,000	698.000
11	COMP B	7.6	8.90	−76.20	80	70	13.800	94,000	315.000
BLDG B-46: 12	A-5	5.1	4.70	−50.80	75	82	0.820	−4,890	0.112
13	A-5	5.1	5.80	−50.80	80	63	0.660	−5,170	0.125
14	A-5	5.1	6.30	−50.80	80	63	0.180	+19,600	1.790

*MAXIMUM VALUES MEASURED DURING THE PLANT SAMPLING

o Minimum ignition energies for explosive and pyrotechnic dusts were reported in the range of 0.2 and 8.0 joules. Maximum energy levels calculated from the charge density measurements were all very low. (maximum energy level of 700 μJ). This was an unusually high reading for Building 04M-40. The highest maximum energy level was in Building 1611 at Louisiana AAP which read 3.0 μJ.

o The charge density appears to be approximately proportional to the peak mass flow rate (duct flow rate, Q, times the maximum dust concentration) in the duct.

RECEIVED May 15, 1987

Chapter 19

Ionizing Air for Static Charge Neutralization While Processing Sensitive Materials

B. V. Diercks

Morton Thiokol, Longhorn Division, P.O. Box 1149, Marshall, TX 75671

Ionized air can be safely and effectively utilized for neutralizing static charges which are generated while processing sensitive energetic materials. The Longhorn Division of Morton Thiokol, Inc. has successfully incorporated systems, in which electrically generated ions are used to neutralize charges which accumulate on infrared energy generating compositions consisting of magnesium powder, polytetrafluoroethylene (PTFE) and a binder. Ignitions sporadically occurred as pressed pellets of the composition were removed from the consolidation press. The ionizing air systems enhanced the safety of this and other infrared composition processing operations.

Morton Thiokol, Inc. is the operating contractor of the Government owned facilities at Longhorn Army Ammunition Plant. The plant is physically located in Karnack, Texas. The Longhorn Division is the Government's primary production facility for illuminating ammunition, signals, pyrotechnic simulators (gun flash, artillery burst, hand grenade, etc.) and infrared decoy flares. An electrostatic problem encountered in 1983 while processing infrared flare composition resulted in the utilization of ionizing air for neutralizing static charges while processing these compositions. Although the use of ionized air to date has been limited to infrared compositions, the techniques employed are applicable to any situation wherein the processing of energetic compositions are susceptible to ignition from electrostatic discharge.

Compositions whose products of combustion produce energy in the infrared wave band are generally composed of magnesium powder, polytetrafluoroethylene (PTFE) and a binder. For efficient tactical utilization of the energy developed by the combustion process the composition is normally formed into pellets either by press consolidation or by press extrusion. The process being used at Longhorn at the time the electrostatic problem was encountered was press consolidation. The composition was being consolidated into a pellet

0097-6156/87/0345-0286$06.00/0
© 1987 American Chemical Society

approximately 2"x3"x5" which weighed approximately 600 grams. The consolidation load was applied to the 2"x5" surface. Grooves were pressed into the top and bottom of the pellet by the upper and lower punches. Additionally, a cavity, into which a Safe and Igniting (S&I) device is installed, was formed during the consolidation process by using a side punch. Figure 1 depicts the consolidation press which was being used. As depicted, the punches are all retracted and ready to receive composition. Upon charging the cavity with material the press sequenced by moving the upper and side punches into position. The lower punch then raised, pressing the composition against the die walls and other punches to form the pellet to the final configuration. After a predetermined dwell time, the lower punch relaxed and the other two punches retracted to their load positions. The lower punch then raised to push the consolidated pellet completely above the die (Figure 2). Ignitions occurred when the pellet was physically removed from the lower punch.

Electrostatic charges are generated when surfaces are separated. At least one of the surfaces has to be a poor conductor although both can be. These charges can then be inductively transferred to and delivered by conductors, or can be transferred to conductors in the form of a spark. The Magnesium/PTFE/Binder pellet is a poor or non-conductive material. As the pellet is removed from the die of the press, electrons are stripped from the walls of the steel die and fluted upper punch and accumulate on the pellet surfaces. At the completion of pellet ejection from the die cavity the pellet contains a negative charge on all exposed surfaces (Figure 2). The magnitude of this charge is not the same on each exposed surface. Because the pellet is a poor conductor, the charge can not dissipate through the grounded lower punch. The situation is one in which a charged pellet is being removed by a grounded and conductive press operator. Although investigation of the incidents revealed the actual pellet ignitions resulted from electrostatic discharge between the pellet and the lower punch created by the physical act of separating the pellet from the lower punch (corrected by using a surfactant to improve the conductivity between the lower pellet surface and the punch), the potential for pellet ignition existed through electrostatic discharge between the charged pellet and the press operator.

To remove or neutralize an electrostatic charge from a poor or nonconductive surface, all points on the surface must be physically addressed. This can be accomplished by sparklessly grounding the entire surface and neutralizing the charge or by "washing" the surface with ionized air. The latter is by far the faster and more positive approach. Figure 3 is a generalized depiction of a voltage versus time profile of a charged pellet exposed to the atmosphere and of one exposed to ionized air.

The use of radioactive ionization sources in areas subject to explosion or fire is undesirable because of the potential for area contamination with radioactive material which could be disseminated in the event of an explosion or fire. With proper precautions, however, electrical ionizing systems can be safely and effectively utilized while processing electrostatically sensitive energetic materials. Ions are generated electrically by corona discharge

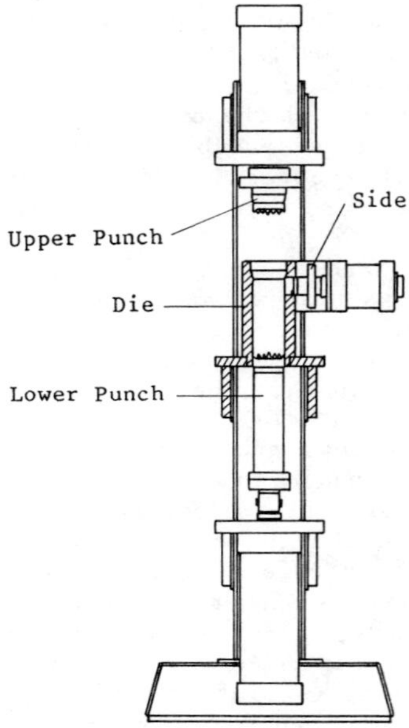

Figure 1. Consolidation Press (Reproduced with permission from Ref. 1. Copyright 1986 Elsevier Science Publishers.)

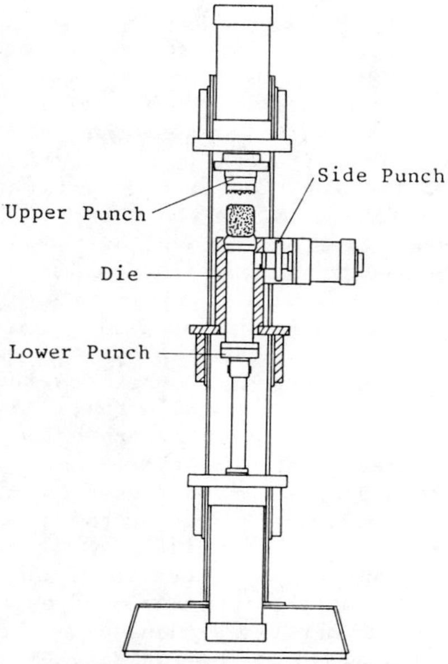

Figure 2. Consolidation Press (Reproduced with permission from Ref. 1. Copyright 1986 Elsevier Science Publishers.)

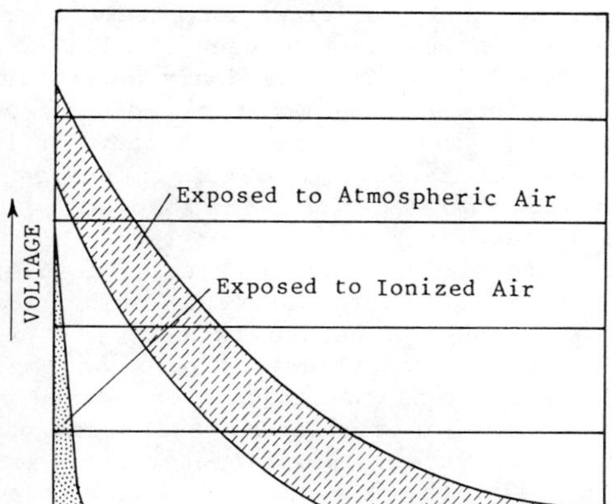

Figure 3. Generalized Voltage/Time Profile

using a needle capacitively coupled to approximately 7500 volts. These corona generators are made in the form of nozzles in which air is forced through the annular space between the high voltage needle and the cylindrical nozzle body. By arranging nozzles in manifolds and coupling them to a common power and air supply, large electrostatically charged surface areas can be effectively neutralized. Such a system was developed for use on the infrared flare pellet consolidation system described above.

Figure 4 depicts the locations and general connections of the nozzle system employed to neutralize the infrared flare pellet as it rests on the lower punch awaiting removal by the operator. The high voltage supply is located in a non-hazardous area and high voltage is cabled to the nozzles. Various interlocks are used to insure that ionized air is proper and present during operations. Contact type interlocks are provided to assure regulated power is delivered to the high voltage power supply. Current to the power supply is metered through relays which have adjustable high and low alarm points. If for any reason the current drops below the low set point or goes above the high set point, the alarm contacts de-energize the press operating controls and the press ceases to function. Furthermore, pressure switches interlocked with the press operating controls (PS# 1,2,3 Figure 4) are installed in the air lines upstream and downstream of the ionizing air nozzles which assures the presence of high pressure air at the ionizing nozzles. Low pressure air is left on the system at all times preventing dust or particulate matter from settling around or on the ionizing electrodes.

Since utilizing the ionizing air system on the infrared flare consolidation press, we have provided for static charge neutralization in other processing areas used for manufacture and handling infrared compositions. Figure 5 shows a horizontal mixer in which flare composition is mixed and masticated. The semi-dry and granular material is dumped from the mixer bowl into a transfer hopper. The dump chute located in the area between the tilted mixer bowl and the open hopper is "washed" with ionized air neutralizing any charge which tends to accumulate on the material as a result of granular attrition. Composition in the transfer hopper is later dispensed into blender buckets for ease of handling in subsequent operations (Figure 6). Material is fed from the hopper by a star valve, through a ring of six ionizing nozzles located in a circular pattern at 60° intervals, into the blender bucket. The blender bucket of material is then dispensed into an oscillating granulator which forces the material through a screen to arrive at a particle size suitable for charging the consolidation dies on the press (Figure 7). The material as it exits the granulator screen falls through a ring of ionizing air nozzles similar to that used on the hopper to blender bucket transfer system. Safety interlocks employed in these systems are essentially the same as used for the consolidation press.

All of the ionizing air systems at Longhorn are located in areas where ultraviolet sensors are used in conjunction with deluge systems for fire protection. Care must be taken to shield the ultraviolet detectors from the ion generating corona source. The systems used at Longhorn are individually shielded with PVC tubing or with hoods.

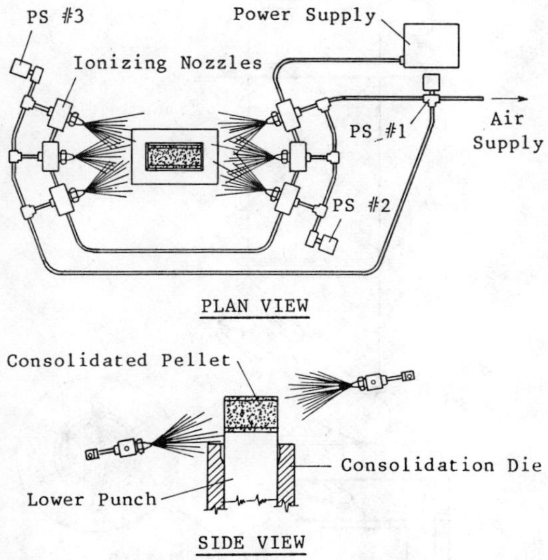

Figure 4. Ionizing Air Nozzles (Reproduced with permission from Ref. 1. Copyright 1986 Elsevier Science Publishers.)

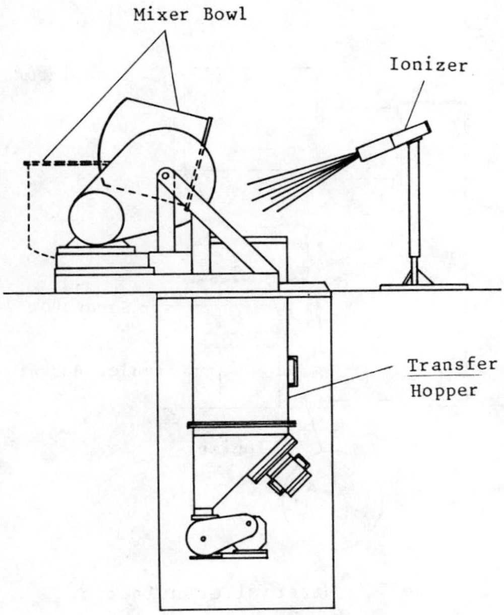

Figure 5. Horizontal Mixer

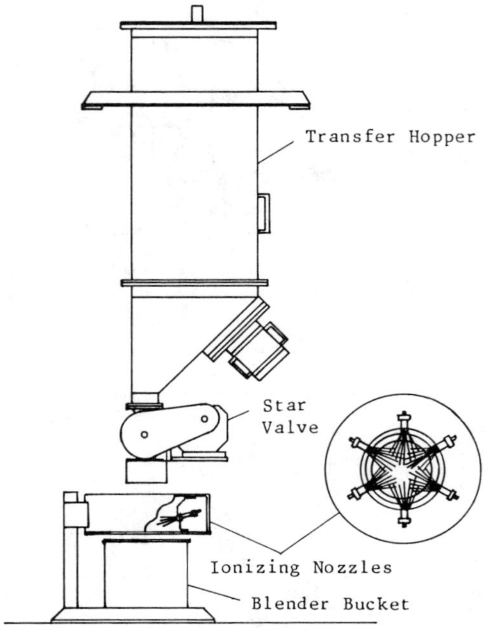

Figure 6. Transfer Hopper to Blender Bucket

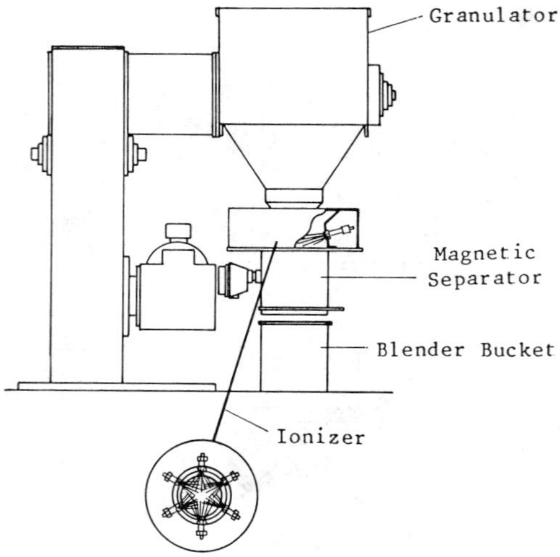

Figure 7. Material Granulation

Ionized air is particularly beneficial in preventing the buildup of electrostatic charges on materials susceptible to the generation of these charges when processing can be readily accomplished in an ionized atmosphere. It is equally effective in quickly neutralizing a charged item or material when the processing environment is not directly accessible with ionized air, but where subsequent processing environments allow it to be subjected to an ionized atmosphere before it must be moved or handled. Caution must be exercised, however, in the installation of these systems to assure that they in themselves do not create a hazardous situation.

Literature Cited

1. Diercks, B. V. Journal of Hazardous Materials, 13 (1986) 3-15.

RECEIVED April 16, 1987

Chapter 20

Design and Use of Ammunition Peculiar Equipment To Protect Workers

Mark M. Zaugg

Tooele Army Depot, Tooele, UT 84074

Discusses the use of Ammunition Peculiar Equipment (APE) used by the military community to perform various operations on ammunition items. Ways in which operators are protected when using APE are included. Specifically, the design of operational shields to contain effects of an explosion is explained.

Ammunition Peculiar Equipment, commonly referred to as APE, is specialized equipment for use in the maintenance, modification, renovation, surveillance and demilitarization of ammunition items. This equipment is used at world wide military installations with ammunition missions that require any of the above mentioned activities.

Whenever the operation to be performed involves the potential to cause the initiation of the propellant, explosive or pyrotechnic (PEP) component(s) of a munition item, the APE is either operated by remote control, with the operator behind a protection wall or barrier, or it is enclosed in a protective barricade or operational shield. Barricades or operational shields are designed to protect personnel and assets from the effects of blast overpressures, thermal effects or fireball, and fragments result from the initiation of PEP components, such as the fuze, primer, propelling charge, burster, etc.

Operational shields are designed and tested in accordance with MIL-STD 398, Shields, Operational for Ammunition Operations, Criteria for Design of and Tests for Acceptance, dated 5 November 1976 (see reference 1). This military standard provides criteria for the protection of personnel and assets from the effects of accidental or intentional detonation and deflagrations, considering the maximum credible incident (MCI) involving the maximum amount of ammunition and explosives within or adjacent to an operational shield, that will detonate or deflagrate as a result of the functioning of a single item.

This chapter not subject to U.S. copyright
Published 1987 American Chemical Society

Operational shields are to be designed to conform to the following requirements:

BLAST ATTENTION. Shields used to provide protection from accidental detonation, are to be designed to prevent exposure of operating personnel to peak positive incident pressures above 2.3 psi or peak positive normal reflected pressure above 5.0 psi.

Shields used to provide protection from intentional detonation of ammunition are to be designed to prevent exposure of operating personnel to impulse noise levels exceeding 140 decibels.

FRAGMENT CONFINEMENT. Shields are to be designed to contain all fragmentation, or direct fragmentation away from areas requiring protection. They are also to prevent generation of secondary fragmentation within areas requiring protection, and prevent movement, overturning, or structural deflections which could result in personnel injury.

THERMAL EFFECTS ATTENUATION. Shield designs are to also limit exposure of personnel to a critical heat flux value based on the total time of exposure. This value of heat flux is determined by the following equation:

$$\phi = 0.62t^{-0.7423}$$

where:

ϕ = heat flux in cal/cm2-sec

t = total time in seconds that a person is exposed to the radiant heat

Operating personnel are to be located at a distance from the shield that assures their exposure is less than the heat flux determined by the above equation. In addition, the upper torso of an operator's body shall not be subjected to any visible fire or flame. Flame impingement upon the lower portion of the body may be permitted provided that the heat flux specified above is not exceeded.

ASSET PROTECTION. Shields intended for intentional detonation are to be designed to prevent damage to buildings, equipment, and other assets in the area. Damage prevention is considered adequate if normal operations are in no way interrupted or hindered as a result of any change to the operational environment from explosions in this type of shield, and the shield may be expected to remain operational throughout its designed life cycle.

Shields designed for accidental explosions only are designed to provide personnel protection from the MCI at that operation and may not, in all cases provide asset protection.

SHIELD DESIGN. In the initial approach to operational shield design, the hydrostatic pressure that would result from the MCI in the shield is determined. For a high explosive detonated in a closed air space, a hydrostatic pressure develops within the space subsequent to the shock wave propagation. This pressure can be found from the equation:

ΔP_o - 4000 hw/v

where:

h = heat of combustion (kcal/gm) (Table I)

w = charge weight (lb)

v = volume of air (ft3)

P_o = Static pressure above ambient (psi)

This equation is derived from the energy equation of state for gas $E = PV(\gamma-1)$, which basically gives the hydrostatic pressure produced by the burning of a substance in a fixed volume of air without a heat loss. (see reference 2). It should be noted that the above relationship applies to bare explosive charges. Static pressure from cased charges will be smaller than those predicted by the equation because of kinetic energy acquired by case fragments. The static pressure decays with time as a function of the heat conduction and convection variables of the shield, and the degree of pressure venting provided.

Table I. Heats of Combustion for Several Explosives are Contained

Explosives	Heat of Combustion kcal/gm
PETN	1.95
RDX	2.28
Pentolite 50/50	2.79
Comp B	2.82
Tetryl	2.93
TNT	3.62
HBX-1	3.73
H-6	3.84
Tritonal 80/20	4.38
HBX-3	4.56

Once the static pressure has been determined, the initial shield design can be done using standard unfired pressure vessel design methods. The geometric shape of the shield is of course driven by the shape of the machine to be contained and the available space in the operating area where the machine and operational shield are to be located. Once the initial design has been made, the dynamic response of the designed shield members to the dynamic pressure is checked. This is necessary to ensure that deflections of structural members due to loading from dynamic pressure produced by the MCI, namely the peak positive incident and reflected pressures, does not permit the escape of fragments or heat flux that would endanger personnel.

Unless specifically designed to do so, operational shields do not totally contain and hold the pressures generated from an explosion. Venting of pressures occurs through joints, flanges, and openings in the shield, and may be enhanced by providing large vented openings that exhaust through the roof or wall of the building in which the shield is located.

The next factor in the shield design is to design for prevention of fragment penetration of the shield material. Fragment penetration can not only be a direct hazard to operating personnel, but partial penetration can weaken the shield causing subsequent failure from the overpressures. Fragment data and criteria for shield design to prevent penetration are contained in chapter 2 of reference 3 and in reference 4.

Knowing that the pressure and fireball within the shield from an MCI will be vented through flanges, openings and joints, the design should provide for long, close tolerance, and if possible, circuitous routes for the pressure and fireball to travel. This will help eliminate passage of fragments outside the shield through openings caused by deflections of shield members. It also provides for quenching of the fireball by heat transfer from the hot gases to the passageway.

SHIELD TESTING.

After the design of the shield has satisfied the requirements, and the prototype shield has been fabricated, reference 1 specifies the testing to which it must be subjected. The prototype operational shield must be tested by creating an MCI in a simulated operational environment.

The MCI is created by detonating or igniting a test round(s), or item(s) with all items in the operational configuration in the shield, including the equipment or reasonable simulation thereof, that performs the intended function on the munitions. If the shield is intended to be used for a variety of rounds, the one(s) having the most severe effects for overpressure, fragmentation, thermal emissions and shape charge effects is to be tested.

For each test the shield must be repaired to the equivalent of new condition or a new shield used, except for shields intended for intentional detonations. Additional explosives equivalent to 25 percent of the explosive filler is added to the test round, if it can be applied in a manner as not to diminish the normal effect and response of the ammunition.

The test should also be conducted in a location that simulates the location when it will be specifically used. For example, shields to be used in an operational bay should be tested in a simulation of an operational bay.

Table II. List of Instrumentation

1.	1 ea	Honeywell 7610 Instrumentation Tape Recorder
2.	1 ea	Artisan EPC 19061 Digital Programmer
3.	4 ea	Kistler 504E Dual Mode Amplifiers
4.	4 ea	Kistler 201B4 Pressure Transducers
5.	1 ea	Medtherm 64 Series Heat Flux Sensor (Schmidt-Boelter type)
6.	1 ea	Systron Donner 8120 Time Code Generator
7.	1 ea	Tektronix 184 Time Mark Generator
8.	7 ea	Honeywell 117 Accudata Amplifiers
9.	1 ea	Krohn-Hite 3202 Variable Filter
10.	1 ea	Honeywell 1858 CRT Visicorder w/1881, 1882 and 1883 Amplifiers
11.	1 ea	ERA TR36-8M Power Supply
12.	1 ea	Newport 60-3 Amplifier
13.	1 ea	HyCam Model 41--0004 High Speed Movie Camera
14.	1 ea	Milliken DSB-5A High Speed Movie Camera
15.	1 ea	Polaroid SX-70 Camera
16.	1 ea	Canon A-1 Reflex Camera
		Support and Calibration Equipment
1.	1 ea	Cohu 335 DC Voltage Standard
2.	1 ea	Dana 5600 Digital Voltmeter
3.	1 ea	Bell & Howell TD 2903-4B Tape Degausser
4.	1 ea	HP 5300A Measuring System
5.	1 ea	HP 3311 Function Generator
6.	1 ea	Beckman 905 WWV Receiver
7.	2 ea	David Clark 10SB-A Sound Powered Head Sets
8.	1 ea	40 ft. Instrumentation Trailer w/installed equipment, racks, patch paneling, lighting systems, heating system, and isolation transformer
9.	1 ea	Kistler 563 A Charge Calibrator
10.	1 ea	Tektronix 561A Oscilloscope with 3A6 Amplifier and 3B4 Time Base Plug-Ins.
11.	1 ea	Pressure Transducer Pulse Calibration Systems

Tests must be properly instrumented to meet the criteria specified earlier in this chapter. All instrumentation should be selected to have the necessary response time and bandwidth equivalent to the anticipated overpressures and heat fluxes. Instrumentation must also be properly calibrated to ensure validity of the data.

Blast pressure gages, heat flux transducers, and sound level meters are to be located at the probable head location of the operator and at representative positions where transient personnel may be located.

Documentation of the tests should also be provided by still photography, video camera/recorder systems, and high speed photography. The high speed photography with a minimum speed of 500 frames per second is necessary to be able to see any flame front exiting a shield. A list of typical instrumentation used on an operational shield test is shown on Table II, (see reference 4).

INDUSTRIAL SAFETY PROVISIONS. In the design of the APE and associated operational shield, conventional machine design practices are used to protect operators from hazards associated with moving parts. Proper techniques for guarding of hazardous machine areas are used, including the use of interlocks in the control system to prevent movements until certain conditions are satisfied, or to stop movements in emergency situations

SUMMARY. The safety record associated with the use of APE operated remotely or within operational shields is excellent. Operational shields that are properly designed, fabricated, and tested do provide operators with adequate protection, and ensures their safety during hazardous operations.

Literature Cited

1. Mil-STD 398; Shields, Operational for Ammunition Operations, Criteria for Design of and Test for Acceptance; 5 November 1976.
2. "Explosives in Enclosed Spaces", U.S. Naval Ordnance Laboratory NavOrd Reports 2934 and 3890; Sixth Symposium on Combustion, Reinhold Publishers, Inc. N.Y., 1957 p. 648
3. CPIA Publication 394, Hazards of Chemical Rockest and Propellants, Volume I, Safety, Health, and the Environment; September 1984, Healy, J., Weissman, S., Werner, H., and Dobbs, N.
4. Priming Fragment Characteristics and Impact Effects on Protective Barriers, Picatinny Arsenal, Dover, New Jersey, Technical Report 4903, December 1975.
5. Miller, J., APE 1011M6 Operational Shield Tests MK39 Primer in Empty 6 Inch/47 Cartridge Case (1340 Grains Black Powder), Ammunition Equipment Directorate, Tooele Army Depot, Tooele, Utah, D/AE Report 05-82, 22 February 1982.

RECEIVED April 21, 1987

Chapter 21

Cleaning Process Lines in the Explosives Industry

Roy W. Wheeler

715 Connie Road, Baraboo, WI 53913

Many hazardous operations require the use of pipelines to convey product material from one location to another. In time, these pipelines become lined with the hazardous product to the extent they could serve as a media for propagation of an incident from station to station. Therefore, periodically cleaning these lines of the hazardous residue is important to operational safety. Additionally, removing accumulations of residue in the pipelines will increase flow volume, operating efficiency, and will minimize the possibility of product contamination.

The necessity to clean these process pipelines varies from desirable to imperative, and the frequency of cleaning may range from weekly to annually or less often.

Even though flanged joints are used to connect sections of pipelines that convey hazardous materials, there is a slight risk of initiating the product when disassembling these connections to gain access to the interior for cleaning the sections.

Circulating a cleaning fluid, or flushing these pipelines with water or fluid is often not effective in removing residual material. The risks can be substantially reduced and residual material can be effectively removed by a method used at an ordnance plant which was placed in an inactive status. After shutdown, thousands of feet of product lines were found to contain hazardous accumulations of residual product, and were thoroughly cleaned in a fraction of the time it would have taken to dismantle these pipelines and clean them by sections. Additionally, the risk of dismantling was practically eliminated.

Cleaning waterlines and fuel pipelines with pipe pigs has been an acceptable practice for many years. However, cleaning pipelines that conveyed explosives with a pipe pig is innovative, and proved to be very effective and economical.

Many varieties of pigs are available, some of which are quite sophisticated. However, very simple pigs are sufficient for most pipe cleaning operations in the explosives industry.

0097-6156/87/0345-0300$06.00/0
© 1987 American Chemical Society

A system can easily be navigated by these pigs, which are propelled hydraulically, at pressures usually substantially less than operating pressures of the system. The pig collects debris and pushes it out of the system and also puts into suspension, material that can be combined into the flow that is propelling the pig.

Pigs may be obtained that are made of metal, rubber or urethanes, or in combination of these materials. The type chosen for use will depend on its compatibility with the product in the pipeline and its durability.

Simply flushing a system, even with full bore flow, and at maximum velocity, is only marginally successful and an unacceptable way to clean many of the pipeline systems in the explosives industry. The carrying ability of the fluid can be relied upon only if the flow can keep everything in suspension or moving. But simple flow cannot loosen and remove encrustations or tuberculation that may be in residence, and that contribute to the possibility of propagation, contamination, or decreased hydraulic capacity. The proper pipe pig is highly successful in accomplishing removal of these obstacles. Pipe pigs can be obtained that negotiate turns and pass through fully opened valves, which eliminates the need to dismantle the pipeline at these locations.

Polyurethane foam pipeline pigs, which were used at the ordnance plant, can be obtained in diameters from 1/2 to 108 inches in increments of 1/8 inch. The most common sizes used at this plant were 8", 10", 12" and 16".

The pipes to be cleaned may be of almost any length. A means of ingress and egress for the pig must be provided. all valves in the line to be cleaned must be fully open. (Valves in any branch line should be kept closed, to insure the pig follows the path of least resistance - the main line.)

Using a pig approximately 1/4 - 1/2 inch larger in diameter than the pipe to be cleaned, the pig is inserted into a larger spool attached to the ingress end of the pipeline. The spool end would then be capped with a plate that has been provided with a fitting to attach the hydraulic line to be used to propel the pig through the pipeline (Figure 1). The ordnance plant used a specially fabricated tapered pipe section that could be attached to the pipeline and be removed after use (Figure 2). Pipelines that require frequent cleaning can be provided with a permanently installed "y" section at the ingress end of the pipe for launching the pig (Figure 3).

The speed of the pig is controlled by regulating the discharge pressure of the hydraulic fluid pressure line. This can be determined and monitored by installing a pressure gauge on the system. The most effective cleaning is obtained when the linear speed of the pig in the pipeline is controlled within 1 to 5 fps (0.3 to 1.5 m/g).

At the egress (discharge) end of the pipeline, provisions should be made to handle the fluid and product being emitted. Explosive products that are insoluble in the hydraulic fluid being used can be discharged into a sump where they can be removed later and destroyed, or through a fine mesh screen that will retain the explosive products for later disposition. Soluble products will require collection and disposition of both the product and the hydraulic fluid.

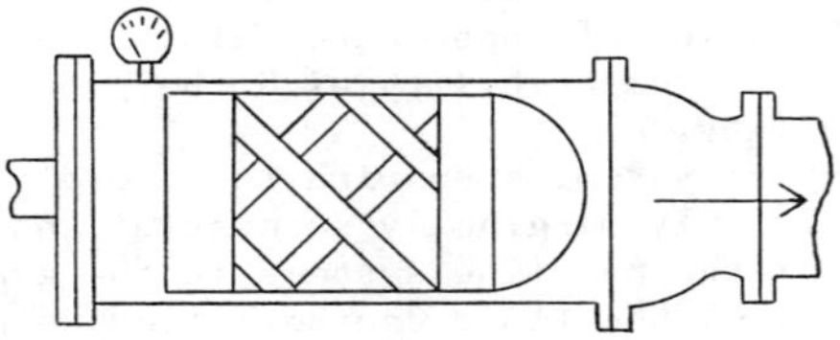

Figure 1. Straight Line Spool Launcher.

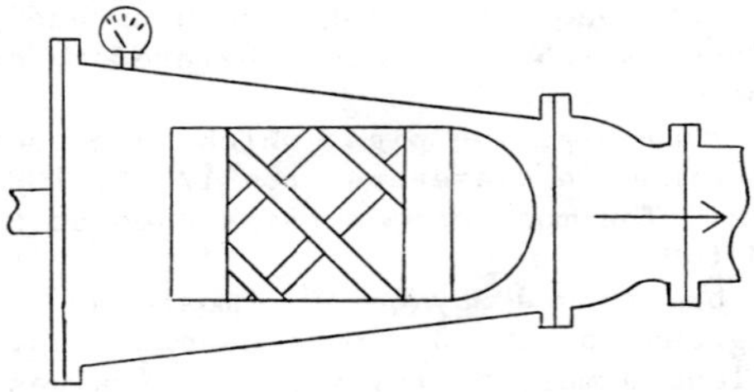

Figure 2. Straight Line Tapered Launcher.

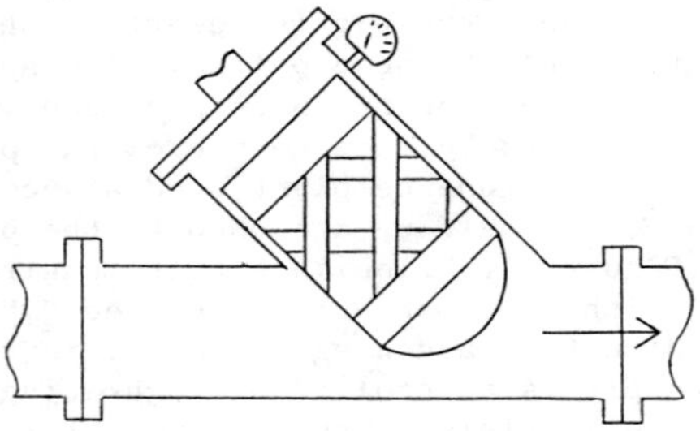

Figure 3. Permanently Installed "Y" Launcher.

It may be desirable to first clean the pipeline with a polyurethane foam swab. This material can be purchased commercially either in specific cut sizes, or in bulk, which can be cut to the desired size. Swabs will effectively remove soft scales and loose material. Their method of use is identical to that of the pig.

While cleaning the pipeline, the swab or pig may encounter a heavy build-up of encrustation, and its progress be interrupted. This would be evidenced by an increase on the pressure gauge. In most cases the swab or pig will progress past the interuption and regain its normal progression. However, if it did not, and the pressure continued to rise without fluctuation, the hydraulic pressure should be allowed to drop and then the pipeline re-pressurized in an attempt to force the pig past the obstacle. In the worst case, where the pig or swab became lodged, it would be necessary to reverse the flow by applying hydraulic pressure on the egress end of the pipeline.

Before adopting this method at the ordnance plant, sections of pipelines were chosen for test samples, to determine if the swab and pig method would satisfactorily clean these contaminated pipes. One half the sections were cleaned by this method and the other half was thoroughly flushed with water. They were allowed to dry and then were subjected to initiation by fires. The sections that had been flushed with water ignited and burned vigorously. The sections that had been subjected to cleaning with the swab and pig had no product remaining that would support combustion.

In keeping with the cardinal principal of safety in the explosives industry, cleaning product pipelines by the pig method exposes personnel to the least amount of hazardous material for the shortest period of time and reduces potentially hazardous disassembly operations to the minimum.

Every explosive operation that requires conveying hazardous material by enclosed pipelines should be considered a candidate for cleaning the pipes by this method.

RECEIVED March 6, 1987

INDEXES

Author Index

Affiliation Index

Subject Index

F

M

N

O

P

Production by Barbara J. Libengood
Indexing by Karen L. McCeney
Jacket design by Carla L. Clemens

Elements typeset by Hot Type Ltd., Washington, DC
Printed and bound by Maple Press Co., York, PA
Dust jackets printed by Atlantic Research Corporation, Alexandria, VA

Recent ACS Books

Personal Computers for Scientists: A Byte at a Time
By Glenn I. Ouchi
276 pp; clothbound; ISBN 0-8412-1000-4

The ACS Style Guide: A Manual for Authors and Editors
Edited by Janet S. Dodd
264 pp; clothbound; ISBN 0-8412-0917-0

Silent Spring Revisited
Edited by Gino J. Marco, Robert M. Hollingworth, and William Durham
214 pp; clothbound; ISBN 0-8412-0980-4

Chemical Demonstrations: A Sourcebook for Teachers
By Lee R. Summerlin and James L. Ealy, Jr.
192 pp; spiral bound; ISBN 0-8412-0923-5

Phosphorus Chemistry in Everyday Living, Second Edition
By Arthur D. F. Toy and Edward N. Walsh
362 pp; clothbound; ISBN 0-8412-1002-0

Pharmacokinetics: Processes and Mathematics
By Peter G. Welling
ACS Monograph 185; 290 pp; ISBN 0-8412-0967-7

Light-Activated Pesticides
Edited by James R. Heitz and Kelsey R. Downum
ACS Symposium Series 339; 339 pp; ISBN 0-8412-1026-8

Solving Hazardous Waste Problems: Learning from Dioxins
Edited by Jurgen H. Exner
ACS Symposium Series 338; 397 pp; ISBN 0-8412-1025-X

Crystallographically Ordered Polymers
Edited by Daniel J. Sandman
ACS Symposium Series 337; 287 pp; ISBN 0-8412-1023-3

Sources and Fates of Aquatic Pollutants
Edited by Ronald A. Hites and S. J. Eisenreich
Advances in Chemistry Series 216; 558 pp; ISBN 0-8412-0983-9

Nucleophilicity
Edited by J. Milton Harris and Samuel P. McManus
Advances in Chemistry Series 215; 494 pp; ISBN 0-8412-0952-9

For further information and a free catalog of ACS books, contact:
American Chemical Society
Distribution Office, Department 225
1155 16th Street, NW, Washington, DC 20036
Telephone 800-227-5558